Sophie Goldie

Hodder Education, 338 Euston Road, London NW1 3BH

Hodder Education is an Hachette UK company

First published in UK 2011 by Hodder Education

This edition published 2011

Artworks (internal and cover): Peter Lubach

Cover concept design: Two Associates

British Library Cataloguing in Publication Data: a catalogue record for this title is available from the British Library.

10 9 8 7 6 5 4 3 2 1

www.hoddereducation.co.uk

Typeset by Stephen Rowling/Springworks

Printed in Spain

'Maths is like love: a simple idea but it can get complicated.'

R. Drabek

About the author

Sophie Goldie has always loved mathematics and has a strong interest in maths education. Having graduated from Warwick University with a BSc in mathematics and physics, she went on to teach mathematics at sixth form level. She has co-authored many textbooks, revision guides and teachers' guides, and has written material for a number of mathematics websites. She is also an A-level assistant examiner.

Acknowledgements

I would like to thank my husband, Scott, for his unfailing support and helpful suggestions. Thanks also to my children, Catherine, Isobel and Peter, for providing a constant stream of welcome distractions.

Sophie Goldie, Alton, Hampshire
June 2011

Contents

Introduction

Maths can seem like a bewildering list of complicated formulae and rules, but this book gets **straight to the heart of the subject**, giving you the lowdown on the **essential points**. This isn't a book that goes into lengthy explanations of why a particular method works or where a formula comes from. Rather, it gives you **handy short-cut methods** as well as **tips for remembering rules and techniques**. In addition to this, there are the formal definitions and facts needed to give you a good grounding in mathematics.

Solving problems is at the heart of mathematics. We all like solving puzzles and I hope that the **friendly, informal style** of this book will help you remember that maths is basically a series of puzzles which need solving!

Maths is by nature a hierarchical subject; however, this is intended as a 'dip-in' book. Each chapter covers the **main methods** that you need to tackle mathematics problems, from dealing with fractions to solving equations. You might find it helpful to read this book with a pen to hand, as maths is definitely a subject to 'do' rather than read. **Have a go at the examples** before reading on to find the solution.

Good luck and happy reading!

Solving problems is at the heart of mathematics

1 Numbers

Some useful words

Our number system is called the decimal system because we count in tens (also called base 10)

- 10 is **divisible** by 5 because you can divide 10 by 5 exactly.
- A **factor** is a number which divides exactly into another number. The **factors** of 12 are 1, 2, 3, 4, 6 and 12 because all of these numbers divide exactly into 12.
- **Integers** are whole numbers, they can be positive or negative.
- **Multiples** are the answers to the times tables. So the multiples of 6 are 6, 12, 18, 24, 30, 36, 42, and so on.

- A **prime number** has exactly two factors, namely 1 and itself. The first ten prime numbers are 2, 3, 5, 7, 11, 13, 17, 19, 23 and 29; 2 is the only even prime number.
- **Product** means the result of multiplying. The product of 3 and 4 is 12.
- **Quotient** means the result of dividing one number by another.
- The **reciprocal** of a number is '1 divided by that number'. So the reciprocal of 4 is $\frac{1}{4}$ and the reciprocal of $\frac{1}{2}$ is 2.
- **Sum** means add. The sum of 3 and 4 is 7.
- **Difference** means subtract.

Calculations

What is $4 + 5 \times 2$? Do you add first? Or multiply?

It would be very confusing if both 14 and 18 were the right answers, so mathematicians have decided on an order in which operations (+, −, × and ÷) should be carried out. The acronym **BIDMAS** helps you remember the right order:

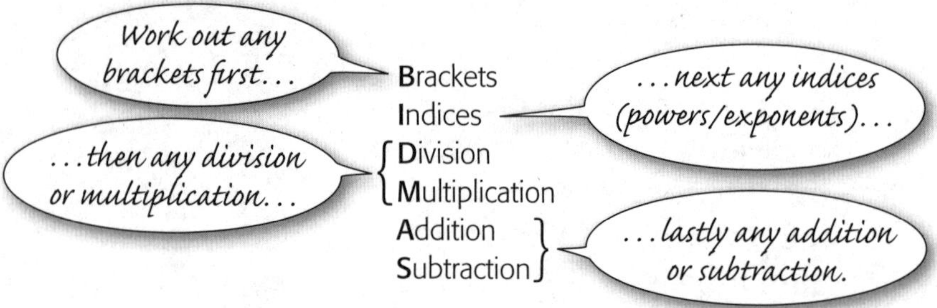

Work out any brackets first... — **B**rackets

...next any indices (powers/exponents)... — **I**ndices

...then any division or multiplication... — **D**ivision, **M**ultiplication

...lastly any addition or subtraction. — **A**ddition, **S**ubtraction

So $7 + 5 \times (4 - 1)^2 = 7 + 5 \times 3^2$ ...brackets first

$= 7 + 5 \times 9$ ...then powers, 3^2 means 3×3

$= 7 + 45$ ...then multiply

$= 52$ ...finally add

To multiply decimals...

...ignore the decimal point and multiply the numbers. Then find the **total** number of **decimal places (d.p.)** in the question – this is the same as the number of decimal places in the answer.

$0.5 \times 0.7 = 0.35$ and $0.4 \times 0.03 = 0.012$

4 × 3 = 12
1 d.p. + 2 d.p. = 3 d.p.

To divide decimals...

...multiply **both** numbers by 10 or 100, etc., to scale the calculation up.

× 10: $5 \div 0.2 = 50 \div 2 = 25$ and × 100: $0.45 \div 0.1 = 45 \div 10 = 4.5$

Directed numbers

- To **add a negative** OR **subtract a positive, move** to the **LEFT** on the number line.

  $2 + (-5) = -3$ is the same as $2 - 5 = -3$

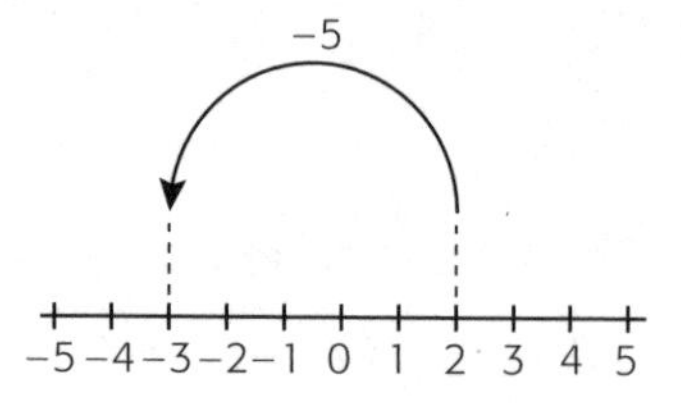

- To **subtract a negative** OR **add a positive, move** to the **RIGHT** on the number line.

  $1 - (-2) = 3$ is the same as $1 + 2 = 3$

- To multiply or divide using negative numbers, ignore signs and carry out the '×' or '÷', then use the following rule:

  » **same signs** means the answer is +
  » **different signs** means the answer is –

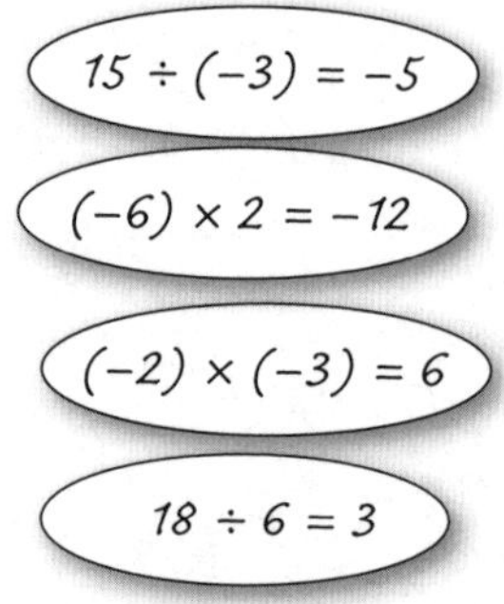

Rounding

TOP TIP

The rule for rounding is look at the **digit to the right** of the one you need to round. If it is:

5 or more, round up

4 or less, round down

3,476 to the nearest 100 is 3,500

18,099 to the nearest 1,000 is 18,000

8.247 to 1 d.p. is 8.2

12.069 to 2 d.p. is 12.07

The first **significant figure (s.f.)** is the first non-zero digit in a number…

346 to 2 s.f. is 350

421,912 to 3 s.f. is 422,000

0.057423 to 2 s.f. is 0.057

0.000832 to 1 s.f. is 0.0008

Factors, multiples and primes

The **least common multiple (l.c.m.)** of two numbers is the lowest number which is a multiple of both numbers.

- Multiples of 4 are 4, 8, **12**, 16, 20, 24, etc.
- Multiples of 6 are 6, **12**, 18, 24, 30, etc.

The lowest number that is on both lists is 12

So the l.c.m. of 4 and 6 is 12.

The **highest common factor (h.c.f.)** of two numbers is the highest number which is a factor of both numbers.

- Factors of 18 are 1, 2, 3, **6**, 9 and 18
- Factors of 12 are 1, 2, 3, 4, **6** and 12

Look for pairs of numbers that × to make 12, e.g. 1 × 12, 2 × 6 and 3 × 4

So the h.c.f. of 12 and 18 is 6.

You can write any number as a **product of prime factors** (this means find prime numbers which multiply to make a given number).

For example:

$$60 = \underbrace{10}_{2 \times 5} \times \underbrace{6}_{2 \times 3}$$

$$= 2 \times 5 \times 2 \times 3$$

It doesn't matter how you start…

$$60 = 5 \times 12$$

$$= 5 \times 3 \times 4$$

$$= 5 \times 3 \times 2 \times 2$$

…you still get the same answer.

So 60 as a product of prime factors is $2 \times 2 \times 3 \times 5$ or $2^2 \times 3 \times 5$.

TOP TIP

Find a pair of numbers which multiply to give 60. Then break those numbers down in the same way until you end up with prime numbers.

Number patterns

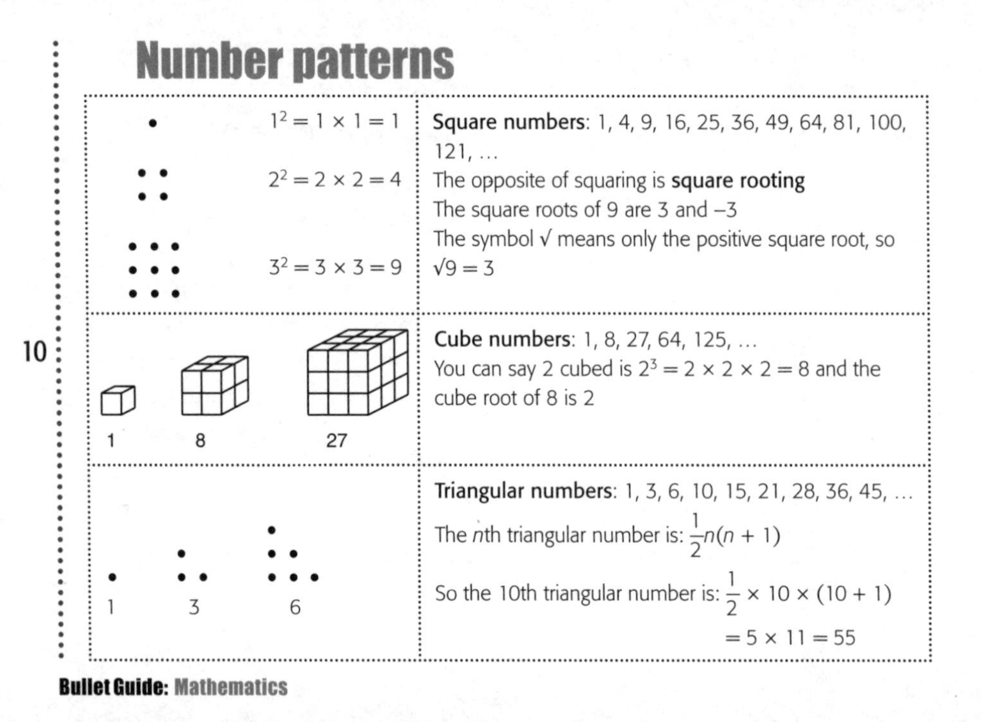

$1^2 = 1 \times 1 = 1$

$2^2 = 2 \times 2 = 4$

$3^2 = 3 \times 3 = 9$

Square numbers: 1, 4, 9, 16, 25, 36, 49, 64, 81, 100, 121, …

The opposite of squaring is **square rooting**

The square roots of 9 are 3 and –3

The symbol √ means only the positive square root, so $\sqrt{9} = 3$

1 8 27

Cube numbers: 1, 8, 27, 64, 125, …

You can say 2 cubed is $2^3 = 2 \times 2 \times 2 = 8$ and the cube root of 8 is 2

1 3 6

Triangular numbers: 1, 3, 6, 10, 15, 21, 28, 36, 45, …

The *n*th triangular number is: $\frac{1}{2}n(n + 1)$

So the 10th triangular number is: $\frac{1}{2} \times 10 \times (10 + 1)$

$= 5 \times 11 = 55$

Laws of indices

a^n means $\underbrace{a \times a \times \ldots \times a}_{n \text{ times}}$	10^4 is $10 \times 10 \times 10 \times 10$
$a^m \times a^n = a^{m+n}$	$9^5 \times 9^3 = 9^{5+3} = 9^8$
$\frac{a^m}{a^n} = a^m \div a^n = a^{m-n}$	$2^7 \div 2^3 = 2^{7-3} = 2^4$
$(a^m)^n = a^{m \times n}$	$(8^3)^4 = 8^{3\times4} = 8^{12}$
$a^0 = 1$ and $a^1 = a$	$5^0 = 1$ and $7^1 = 7$
$a^{-n} = \frac{1}{a^n}$	$10^{-3} = \frac{1}{10^3} = \frac{1}{1000}$

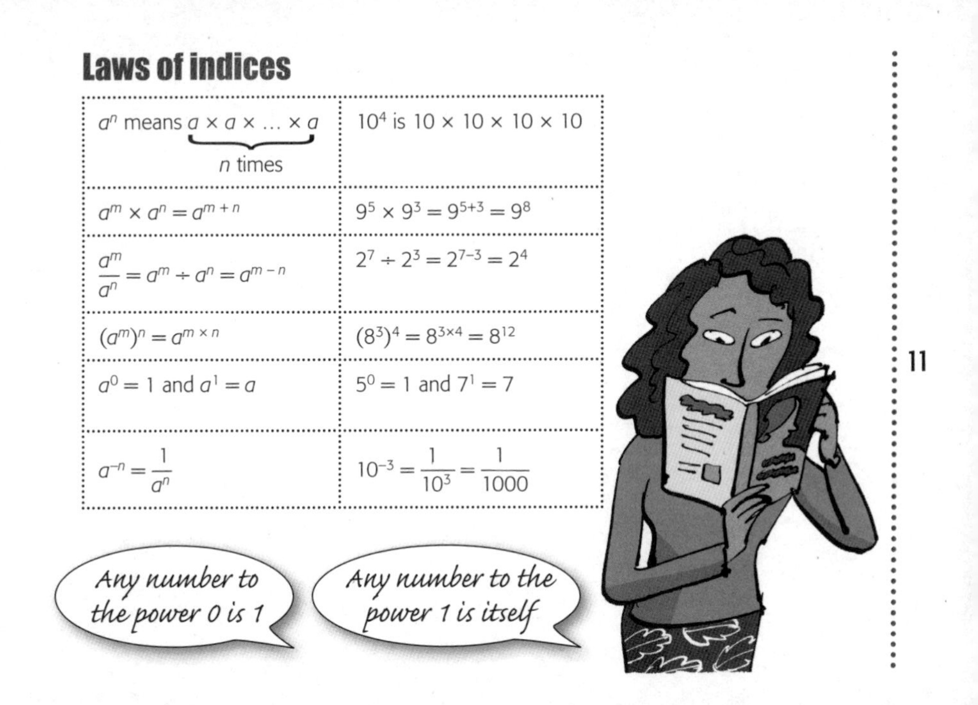

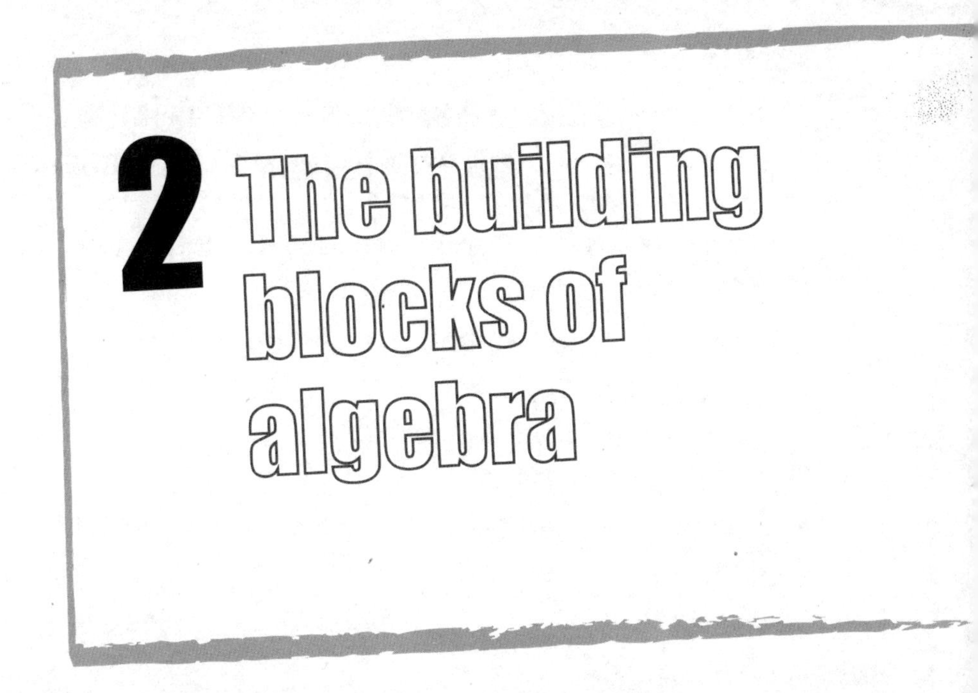

2 The building blocks of algebra

Symbols

In algebra, symbols – usually letters – are used to represent numbers

Symbols such as x, y, θ and π are used as shorthand – they make life easier!

You can use a letter symbol to represent:

- ✔ *any number (a **variable**)*
- ✔ *a particular number like pi, $\pi = 3.14\ldots$ (a **constant**)*
- ✔ *an unknown number that you want to find.*

You cannot use a symbol to represent a word; it must be a number:

a = apples ✘ a = number of apples ✔

- You can use symbols, numbers and the operations '×' and '÷' to make **terms**.
- You can join terms using the operations '+' and '–' to make **expressions**.
- You can connect two expressions with an '=' sign to make **equations** and **formulae**.
- A **formula** is a rule used to work something out.

$A = \pi r^2$ $E = mc^2$

- An **equation** is a statement which says: 'this expression = that expression'.

$2n + 1 = 3n + 4$ $x + y = 10$

Rules for writing algebra

- Don't use a '×' symbol or a '÷' symbol:
  write ab not $a \times b$
  write $\frac{a}{b}$ not $a \div b$
- Write numbers first, then letters:
  $4n$ ✔ $n4$ ✘
- In general, write letters in alphabetical order:
  write xyz not zxy

Letter symbols can be used to represent any number: integers (whole numbers), negative numbers, fractions or decimals.

Expression	What does it mean?
$4n$	$4 \times n$ **or** 4 lots of n **or** $n + n + n + n$
$\frac{n}{5}$	n divided by 5 **or** $n \div 5$
n^2	n squared **or** $n \times n$
$\sqrt{n}$	positive square root of n
$2(n - 3)$	$2 \times (n - 3)$ **or** $2n - 6$
$(3n)^2$	'$3n$ all squared' **or** $3n \times 3n$
$3n^2$	3 lots of n squared **or** $3 \times n \times n$

Substitution

This means evaluating an expression or formula by replacing the letter symbols with numbers and then working out the resulting calculation.

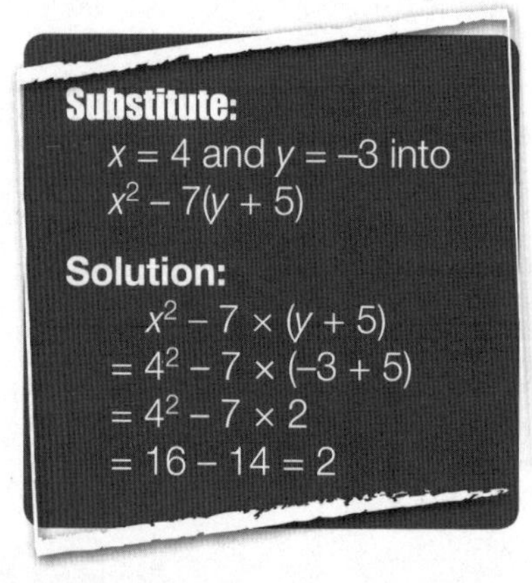

Evaluate:

$A = \frac{1}{2}(a + b) \times h$ when

$h = 7.2$, $a = 4$ and $b = 6$

Solution:

$$A = \frac{1}{2}(a + b) \times h$$
$$= \frac{1}{2}(4 + 6) \times 7.2$$
$$= \frac{1}{2} \times 10 \times 7.2$$
$$= \frac{1}{2} \times 72 = 36$$

Simplifying

Like terms contain the same combination of letters:

*The **coefficients** are the numbers (constants) in front of each term*

✔ *$3xy$ and $5xy$ are **like** terms*
✘ *$4ab$ and $3a^2$ are **unlike** terms*

You can **simplify** an expression by **combining like terms** into a single term. You do this by adding their coefficients – these are the numbers (constants) in front of each term.

Simplify these...

1 $9x - 4 - 7x + 10$
$= 9x - 7x - 4 + 10$
$= 2x + 6$

2 $3x + 5y - 2x + 4y$
$= 3x - 2x + 5y + 4y$
$= x + 9y$

3 $3x^2y + 3y - 4y + 2x^2y$
$= 3x^2y + 2x^2y + 3y - 4y$
$= 5x^2y - y$

Gather like terms first

Brackets

You can remove brackets by **expanding** or **multiplying out**.

Multiply each term inside the bracket by the term outside the bracket. Take extra care with any negative terms. Sometimes you can then simplify the resulting expression.

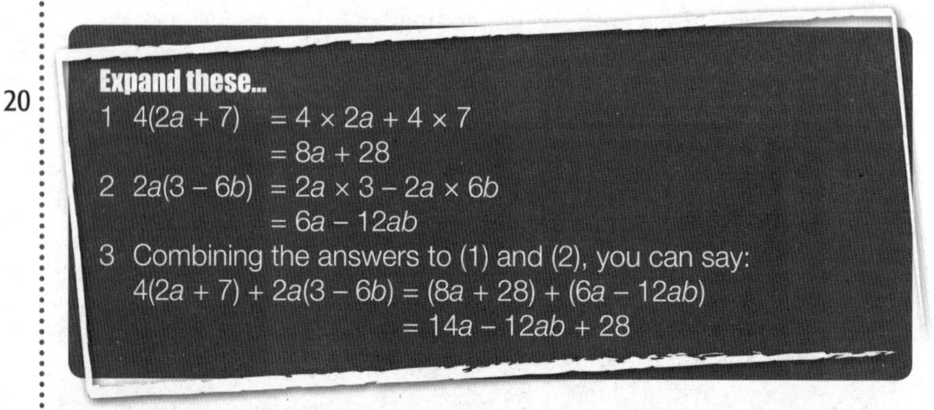

Expand these...

1 $4(2a + 7) = 4 \times 2a + 4 \times 7$

$= 8a + 28$

2 $2a(3 - 6b) = 2a \times 3 - 2a \times 6b$

$= 6a - 12ab$

3 Combining the answers to (1) and (2), you can say:

$4(2a + 7) + 2a(3 - 6b) = (8a + 28) + (6a - 12ab)$

$= 14a - 12ab + 28$

Factorizing

This is the reverse of expanding – you need to rewrite the expression using brackets.

To **factorize fully**, find the highest number and any letters that divide into each term (the highest common factor).

Factorize these...

1. $8a + 12$
   4 goes into both 8 and 12.
   So $8a + 12 = 4 \times 2a + 4 \times 3 = 4(2a + 3)$
2. $15ab - 9a$
   3 goes into both 15 and 9.
   a goes into both ab and a.
   So you can put $3a$ outside the brackets…
   so $15ab - 9a = 3a \times 5b - 3a \times 3 = 3a(5b - 3)$

You can check these are right by expanding the brackets

Sequences

A **sequence** is an ordered list of numbers which follow a rule.

Each number in the list is called a **term**.

Term number	①	②	③	④	⑤	...
Sequence	3,	7,	11,	15,	19,	...

'...' means the sequence carries on in the same way for ever

There are two types of rule you can use for sequences:

1 A **term-to-term** rule
You are given one term (usually the first) and told how to work out the next.

First term is 3
To find the next term, add 4

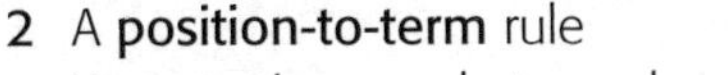

2 A **position-to-term** rule

You are given a rule to work out any term using its term number.

n stands for the term number

nth term is $4n - 1$

1st term is $4 \times \mathbf{1} - 1 = 3$

2nd term is $4 \times \mathbf{2} - 1 = 7$

3rd term is $4 \times \mathbf{3} - 1 = 11$

The difference between each pair of terms is 4

To find the position-to-term rule:

	+4	+4	+4	+4			
Sequence	3,	7,	11,	15,	19,	…	
4 times table	4,	8,	12,	16,	20,	…	$4n$

So the sequence is based on the 4 times table

Each term is **1 less** than the four times table.

So the rule for the nth term is $4n - 1$.

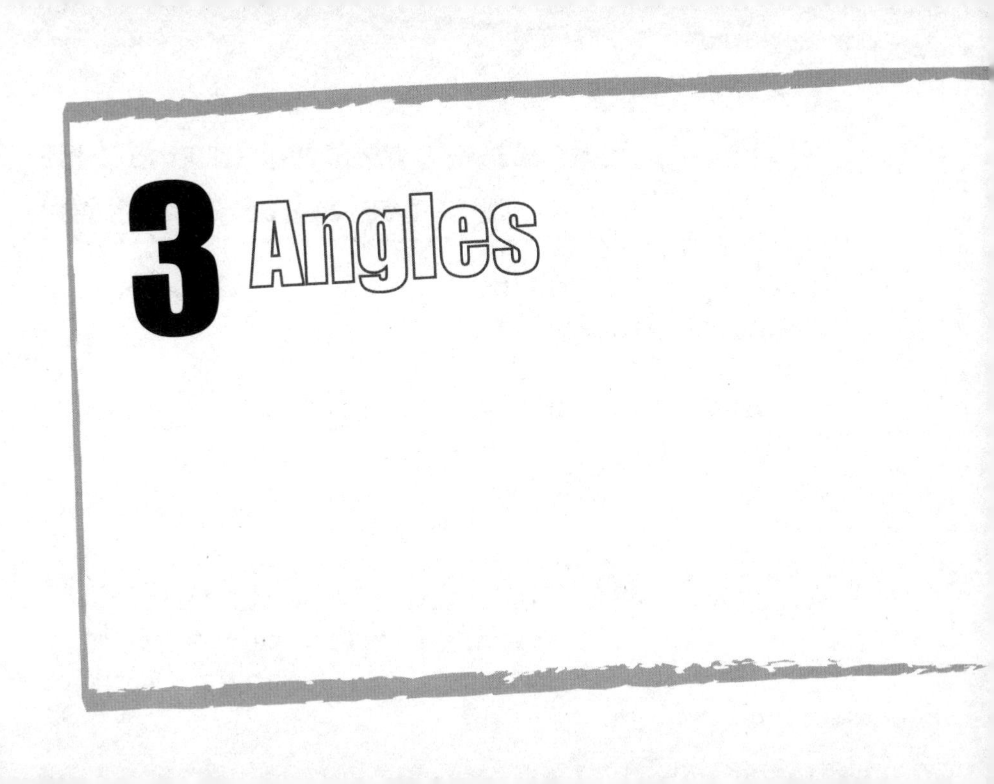
3
Angles

Some useful words

An angle is made when two straight lines (arms) meet at a corner (vertex)

- An angle measures the amount of turn between the two arms.
- Angles are measured in **degrees** (°).

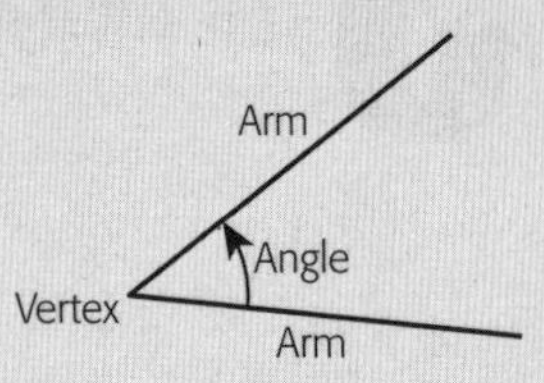

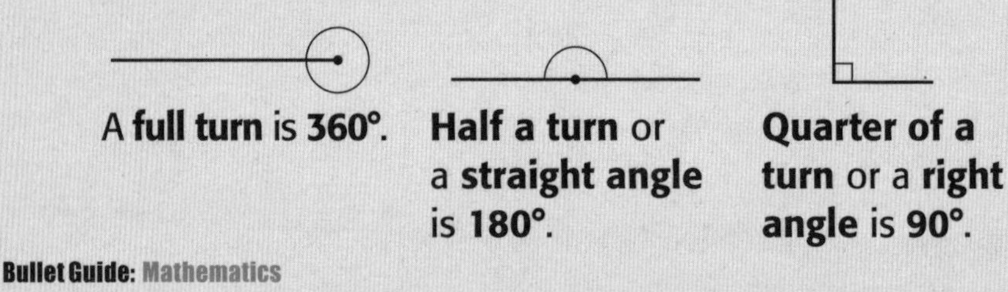

A **full turn** is **360°**.

Half a turn or a **straight angle** is **180°**.

Quarter of a turn or a **right angle** is **90°**.

* **Parallel lines** never meet – they are always the same distance apart (like rail tracks). Arrowheads are used to show parallel lines.
* **Perpendicular lines** meet at right angles.
* An **acute angle** is smaller than a right angle. Think *'a-cute little angle'* to help you remember!

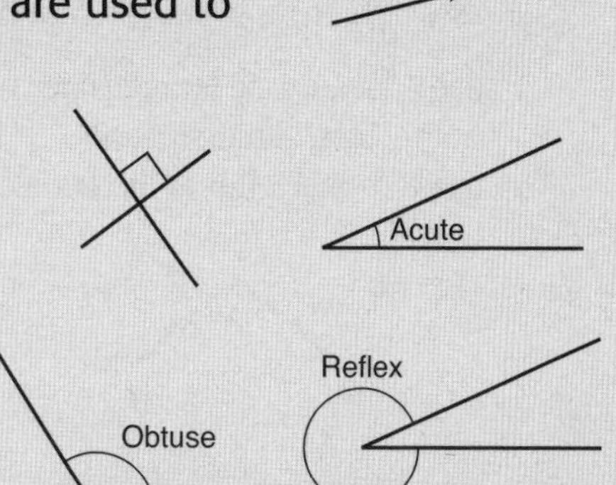

* An **obtuse angle** is larger than a right angle (90°) but smaller than half a turn (180°).
* A **reflex angle** is greater than half a turn.

Angle names

Angles are given names (**labels**) so that you know which angle is being referred to. You can…

1. use a lower case letter inside the angle, like x or the Greek letters α or θ.
2. use a capital letter at the vertex, $\angle$A.

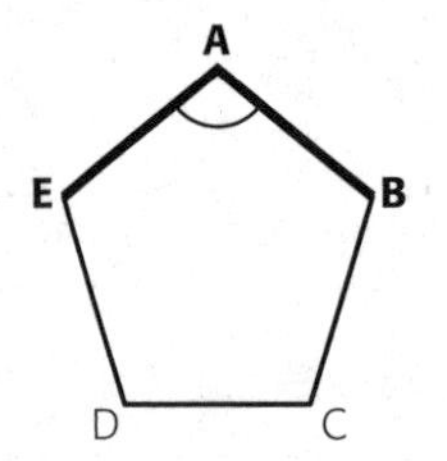

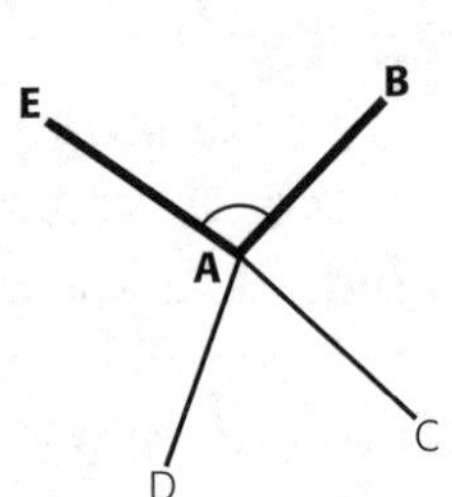

3. use the three letters for the arms and vertex of the angle. The middle letter is always the vertex. So write $\angle$EAB or E$\hat{A}$B.

* A pair of angles that add up to 90° are called **complementary angles**.
* A pair of angles that add up to 180° are called **supplementary angles**.
* When a pair of lines cross, the angles **vertically opposite** each other are equal.

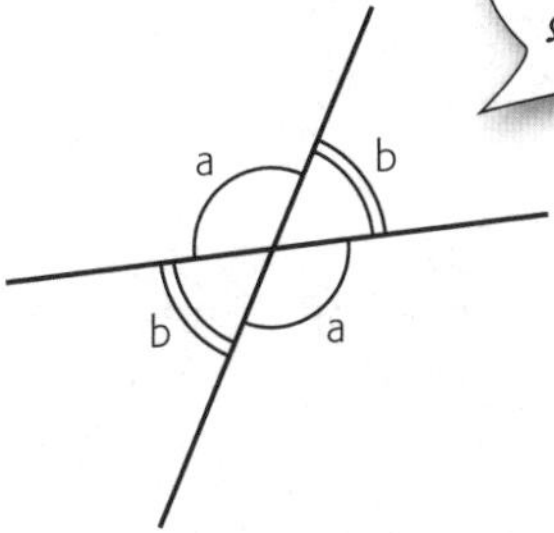

Angle problems

Some angle problems involve a straight line (**transversal**) crossing a pair of parallel lines.

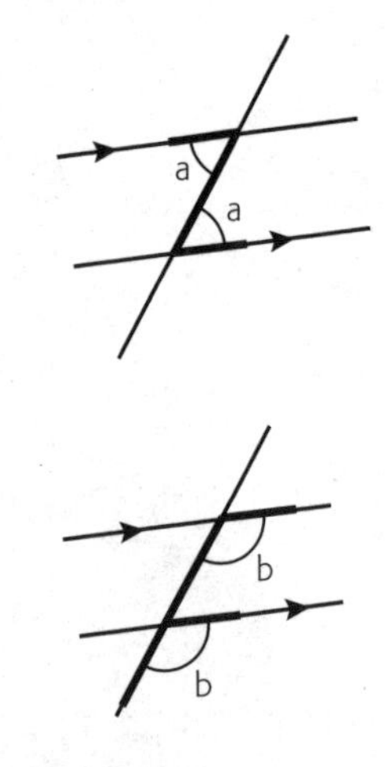

- **Alternate angles** are equal. Look for the *Z* shape – they are also called ***Z*-angles**.

- **Corresponding angles** are equal. Look for the *F* shape – they are also called ***F*-angles**.

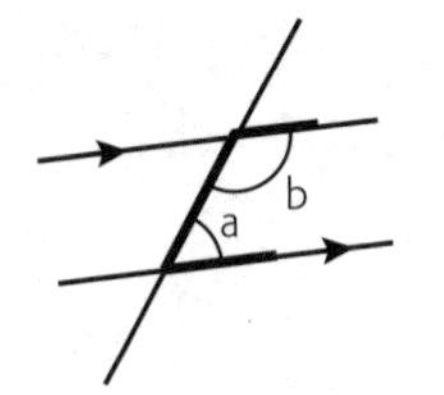

* **Interior angles** add up to 180°, $a + b = 180°$. Look for a ***C*** or ***U*** shape.

It looks like there are lots of different angles to find, but in fact there are only two different angles: one 'small' angle and one 'BIG' angle.

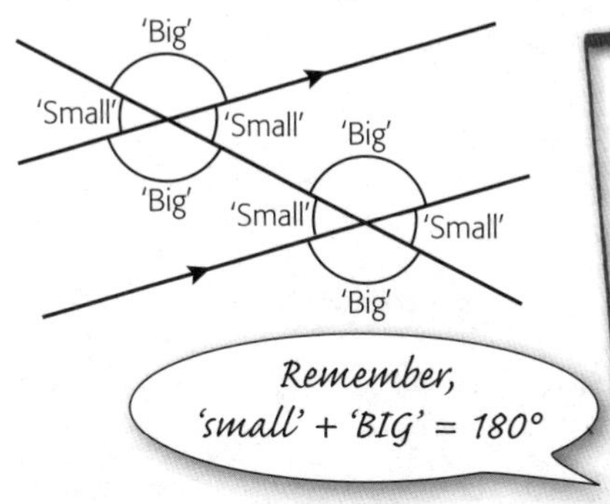

TOP TIP

Once you know one angle, you can work out all the others. So if the 'small' angle = 30°, then the 'BIG' angle = 150° because 30° + 150° = 180°.

Regular polygons

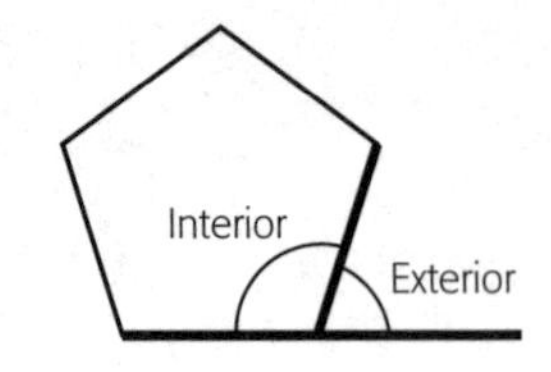

* A polygon is a many-sided, two-dimensional (2D) shape. A **regular polygon** has all sides equal and all angles equal.

* For any polygon, if you walked around the perimeter (outside) you would make a full turn. So the sum of the exterior angles is 360°.

* One exterior angle $= \dfrac{360°}{\text{number of sides}}$

* Interior angle + exterior angle = 180°

* For a polygon with n sides: one interior angle $= 180° - \dfrac{360°}{n}$

  and sum of interior angles $= (n - 2) \times 180°$

* A polygon **tessellates** (can tile a floor without leaving any gaps) when its interior angle divides exactly into 360°.

Tessellations

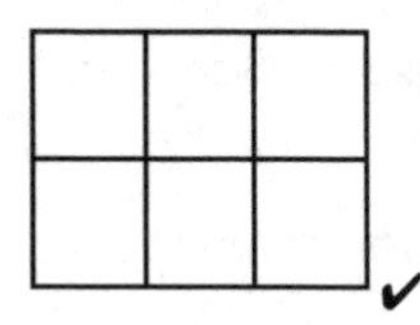

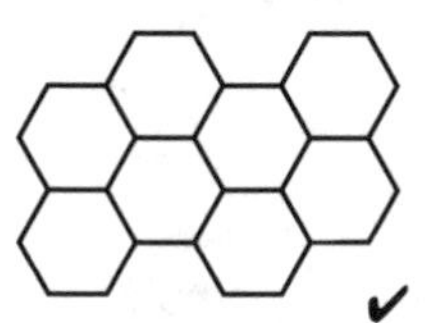

Some facts about polygons…

Regular shape	Number of sides	Exterior angle	Interior angle	Angle sum
Equilateral triangle	3	120°	60°	180°
Square	4	90°	90°	360°
Pentagon	5	72°	108°	540°
Hexagon	6	60°	120°	720°
Heptagon	7	51.42…°	128.58…°	900°
Octagon	8	45°	135°	1080°
Nonagon	9	40°	140°	1260°
Decagon	10	36°	144°	1440°

Bearings

There are lots of ways to help you remember the compass points **N**orth, **E**ast, **S**outh and **W**est: '**N**ever **E**at **S**hredded **W**heat' is one example.

Bearings are used in navigation for direction. They are measured in a clockwise direction from North in degrees and written using three figures. Always draw a **North line** when measuring or drawing bearings. Due North is on the bearing 000° and due East is 045°.

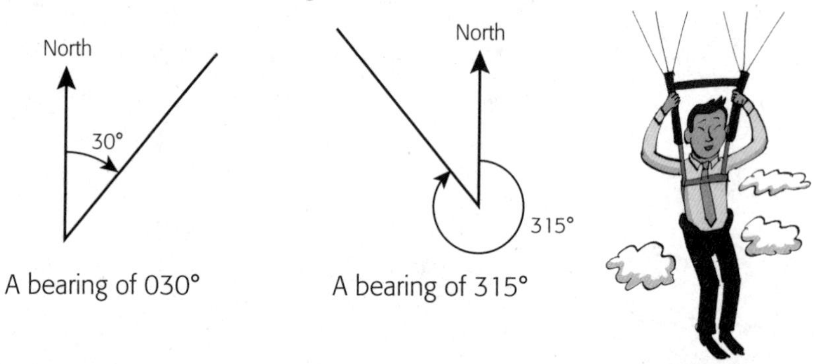

A bearing of 030°   A bearing of 315°

'The bearing of A **from B**' means **measure** the bearing **at B**.

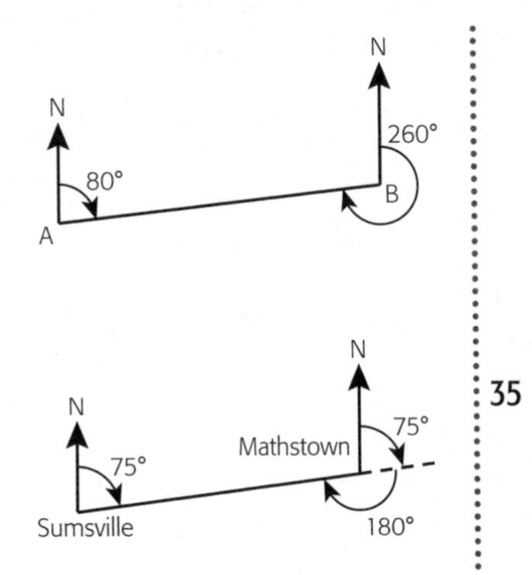

- The bearing of A **from B** is 260°.
- The bearing of B **from A** is 080°.

The **back bearing** is the bearing that takes you back to where you started.

Example

The bearing of Mathstown from Sumsville is 075°. What is the bearing of Sumsville from Mathstown?

Solution: The back bearing is $75° + 180° = 255°$.

4
Fractions and
percentages

Some useful words

You can use fractions and percentages to express proportions

The **proportion** of black or white circles can be described as a **fraction**:

$\frac{2}{5}$ are black, $\frac{3}{5}$ are white

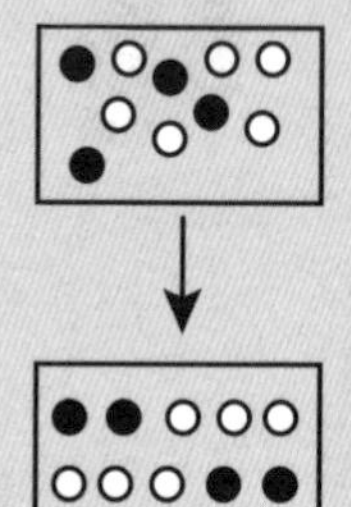

...and as a **percentage**:

40% are black, 60% are white

There are two black circles for every three white circles so:

* the ratio of black circles to white circles is 2:3
* the ratio of white circles to black circles is 3:2

A **fraction** is a number like $\frac{3}{4}$

You can think of a fraction as being:

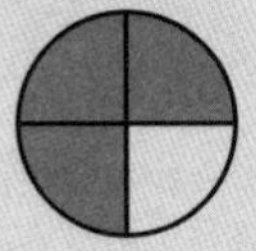

* the number of parts of a whole *or*
* as a number on a number line.

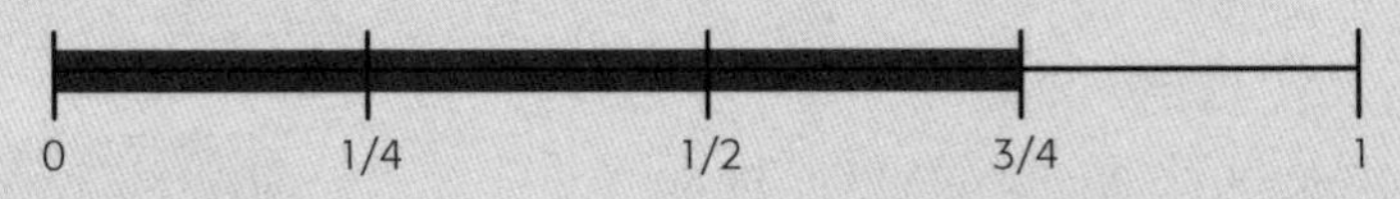

* The **numerator** is the top number of a fraction.
* The **denominator** is the bottom number of a fraction.
* A **simple** or **vulgar fraction** is a fraction where both the 'top' and 'bottom' are whole numbers.

Fractions and decimals

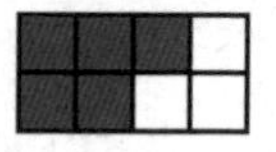

- A fraction like $\frac{5}{8}$, where the 'top' number is less than the 'bottom', is called a **proper fraction**. Proper fractions are less than 1.

- In an **improper fraction** like $\frac{5}{3}$ the 'top' number is greater than the 'bottom'. Improper fractions are greater than 1.

- You can write improper fractions as **mixed numbers** or vice versa.

$$\frac{5}{3} = 1\frac{2}{3} \text{ and } 2\frac{1}{5} = \frac{11}{5}$$

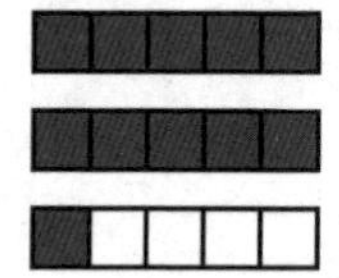

- You can find the decimal value of a fraction by working out 'top' ÷ 'bottom'.

  $\frac{3}{4} = 3 \div 4 = 0.75$ and $\frac{5}{8} = 5 \div 8 = 0.625$

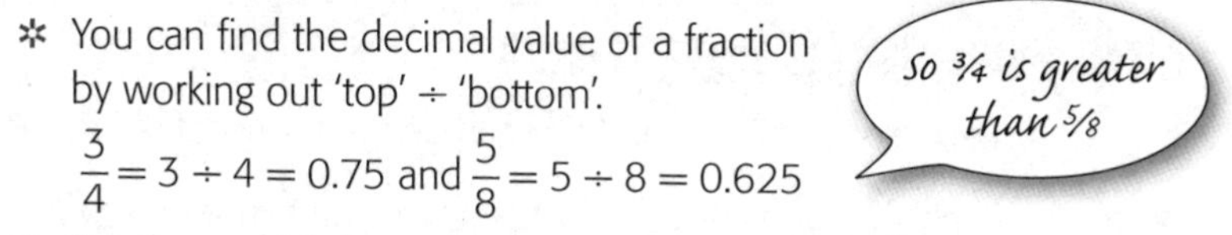

- Fractions which have the same value are **equivalent**. You can find equivalent fractions by multiplying or dividing the top *and* bottom of a fraction by the same number.

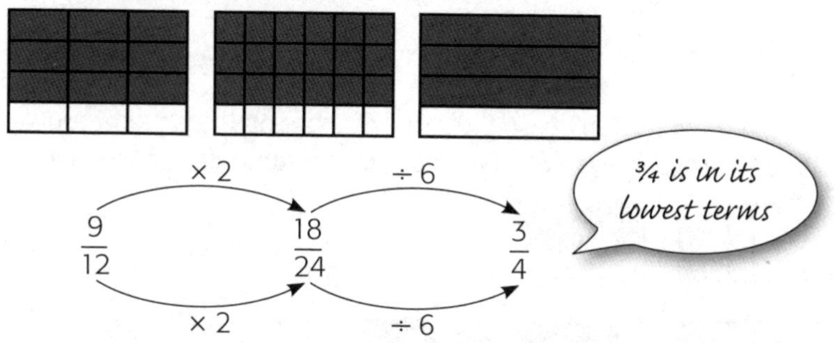

When there is no number that divides into the top and bottom, the fraction is **fully simplified** or in its **lowest terms**.

Working with fractions

* Fractions with **common denominators** have the same bottom numbers.
* You can only add and subtract fractions with the **same** denominators.

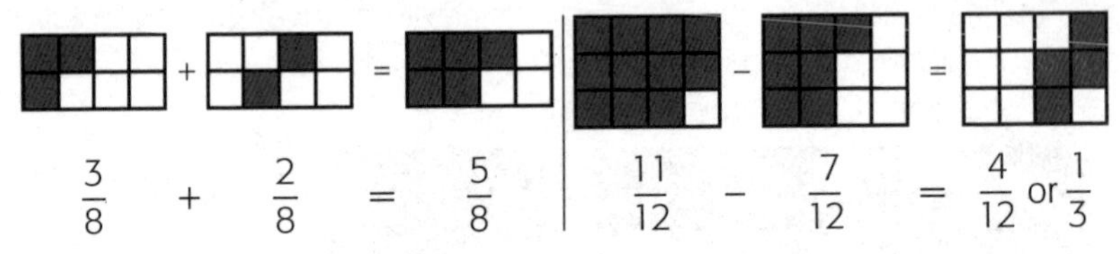

* When fractions don't have the same number on the bottom, you need to find **equivalent fractions** with the **same denominators**.

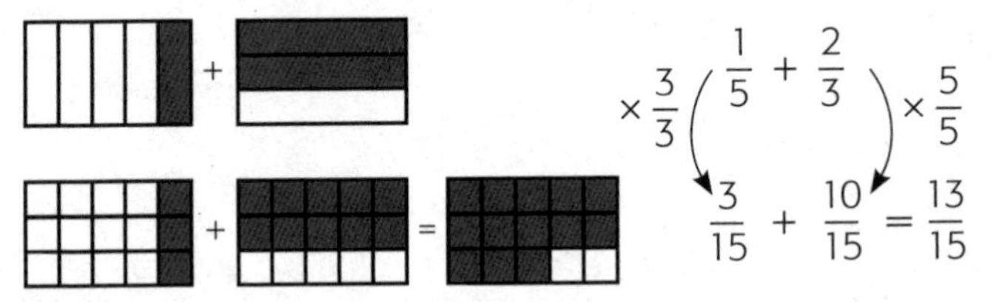

* In maths, 'of' means '×'. So $\frac{2}{3} \times 18$, $18 \times \frac{2}{3}$ and $\frac{2}{3}$ of 18 all mean the same.

  $\frac{1}{3}$ of $18 = 18 \div 3 = 6$, so $\frac{2}{3}$ of $18 = 2 \times 6 = 12$

* To multiply two fractions together:

$$\frac{3}{4} \times \frac{2}{5} = \frac{6}{20}$$
$$= \frac{3}{10}$$

Multiply the tops...

...and the bottoms...

...then simplify.

* To divide by a fraction:

$$\frac{5}{7} \div \frac{3}{4} = \frac{5}{7} \times \frac{4}{3}$$
$$= \frac{20}{21}$$

Turn the 2nd fraction upside down, $\frac{3}{4} \rightarrow \frac{4}{3}$

...and then multiply.

Percentages

✲ **Per cent** (%) means per 100.
To write a percentage as a decimal, divide by 100.

100% = 1 75% = 0.75 or $\frac{3}{4}$
10% = 0.1 or $\frac{1}{10}$ 1% = 0.01 or $\frac{1}{100}$

TOP TIP
Remember, 'out of' means divide.

✲ To convert between fractions/decimals and percentages:

13 out of 20 as a fraction is $\frac{13}{20}$
as a decimal is $13 \div 20 = 0.65$
as a percentage is $0.65 \times 100\% = 65\%$

✲ To find a percentage of an amount: e.g. 72% of £30
Write the % as a decimal… = 0.72 × £30
…then multiply = £21.60

- To find percentages without a calculator, you can start with 10%.

  To work out 35% of 80 kg:

  Divide by 10 to find 10%

  10% of 80 kg = 8 kg

  and so...

  30% is 24 kg

  5% is 4 kg

  So 35% of 80 kg is 28 kg

Example: Increase £60 by 10%

Adding on the %:
10% of £60 is £6
£60 + £6 = £66

Using multipliers:
100% + 10% = 110%
110% of £60 = 1.1 × £60 = £66

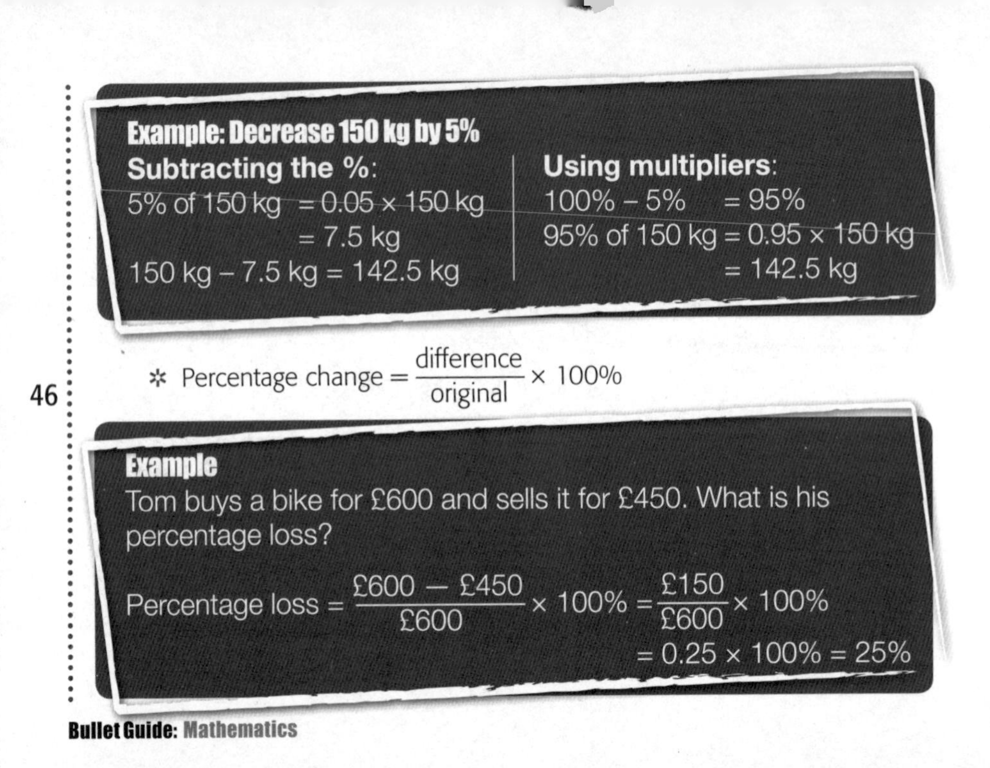

Example: Decrease 150 kg by 5%

Subtracting the %:

5% of 150 kg $= 0.05 \times 150$ kg
$= 7.5$ kg
150 kg − 7.5 kg = 142.5 kg

Using multipliers:

100% − 5% = 95%
95% of 150 kg $= 0.95 \times 150$ kg
$= 142.5$ kg

* Percentage change $= \dfrac{\text{difference}}{\text{original}} \times 100\%$

Example

Tom buys a bike for £600 and sells it for £450. What is his percentage loss?

$$\text{Percentage loss} = \frac{£600 - £450}{£600} \times 100\% = \frac{£150}{£600} \times 100\%$$
$$= 0.25 \times 100\% = 25\%$$

Ratios

Ratios are used to **compare quantities**. So a drink made with 2 parts orange juice to 5 parts lemonade is in the ratio 2:5.

Take care – a ratio of 4:10 or 6:15 would give the same drink, but 5:2 is different!

2 + 5 = 7 parts

1 Divide 14 litres in the ratio 2:5

7 parts = 14 litres

(÷ 7 on both sides)

1 part = 2 litres

So 2 parts orange = 4 litres and 5 parts lemonade = 10 litres.

2 The ratio of men to women in a company is 5:3. There are 60 women. How many men are there?

The company is 3 parts women

3 parts = 60 people

(÷ 3 on both sides)

1 part = 20 people

(× 5 on both sides)

5 parts = 100 people So there are 100 men.

5 Statistics

Collecting data

Statistics is about collecting, presenting and interpreting data to test a hypothesis

A **hypothesis** is a statement that you wish to test using:

* an **experiment** (toast mostly lands 'butter-side' down)
* or a **survey** (people are concerned about global warming).

You need to have a hypothesis *before* you set about collecting data.

When you carry out an experiment, you need to make sure you carry out enough trials.

When you carry out a survey you need to:

* decide on your **population** and whether you are going to survey the whole population (a **census**) or take a **sample**
* avoid **bias** by choosing your sample fairly, so it is representative of the whole population.

The data you collect could be:

* **categorical**: non-numerical data, e.g. eye colour
* **discrete**: numerical data – usually whole numbers
* **continuous**: numerical data which can be any value in a given range, e.g. weights of babies.

Survey example

Hypothesis: employees at company X want to reduce their hours.
Population: all employees of company X.
Sample: choose 50 employees at random.

Tips for writing a good questionnaire

- Don't use **open** questions – for example:
  - What do you think about...?
- Do use questions which need tick box answers (**closed** questions).
- Don't use overlapping categories for tick boxes – for example:

How old are you?	How old are you?
20 – 30 ☐ ✘	20 – 29 ☐ ✔
30 – 40 ☐ ✘	30 – 39 ☐ ✔

- Don't use **biased** or **leading** questions – for example:
  - Don't you agree that… ✘
  - Do you agree or disagree that… ✔
- Do make sure your questions are **relevant** to your hypothesis.

Averages

An average gives a **typical** value for the data. There are three main types of average:

1 **mean**
- » total all the data values
- » divide by number of values.

2 **median**
- » put data in order of size
- » find the middle value.

3 **mode**
- » find the most common (frequent) data value.

Mean
✔ *Uses all the data values*
✘ *Extreme values may make mean unrepresentative*

Median
✔ *Good for skewed data*
✘ *Doesn't take into account spread of data*

Mode
✔ *Can be used for categorical data*
✘ *May be more than one mode or no mode at all*

Range

- The range tells you how spread out the data is.
- Range = highest data value – lowest data value.

Here is the salary for each employee in a company:

£18k, £18k, £26k, £30k, £32k, £33k, £36k, £38k, £120k

Mean $= \dfrac{\text{add up all the values}}{\text{number of values}} = \dfrac{£351\text{k}}{9} = £39\text{k}$

Mode = £18k

Median = £32k

Range = £120k – £18k = £102k

The median is the most representative in this case, as the single high earner has dragged up the mean, and most earn more than the mode.

Frequency tables

You can use a **frequency table** to collect data.

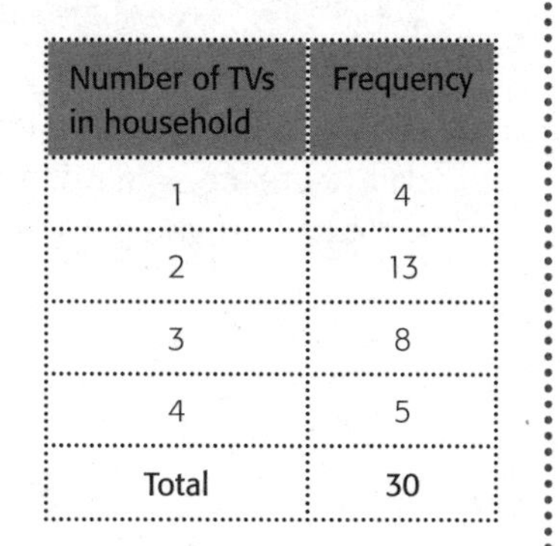

Number of TVs in household	Frequency
1	4
2	13
3	8
4	5
Total	**30**

- The modal number of TVs is 2 (as this has the highest frequency).
- The median is 2 (there are 30 values, so the median is the 15th value. The first 4 values are '1' and the next 13 values are '2' so the 15th value is 2).

$$\text{Mean} = \frac{1 \times 4 + 2 \times 13 + 3 \times 8 + 4 \times 5}{30}$$

$$= \frac{74}{30}$$

$$= 2.47 \text{ to 2 d.p.}$$

Multiply the numbers in each column together (1 × 4, 2 × 13 and so on) to find the total number of TVs.

Presenting data

You can use a pictogram, bar chart or a pie chart to present data. Here are some charts for the data from the previous page:

TOP TIP

Keep bars same width and pictures same size, keep bars/pictures equally spaced.

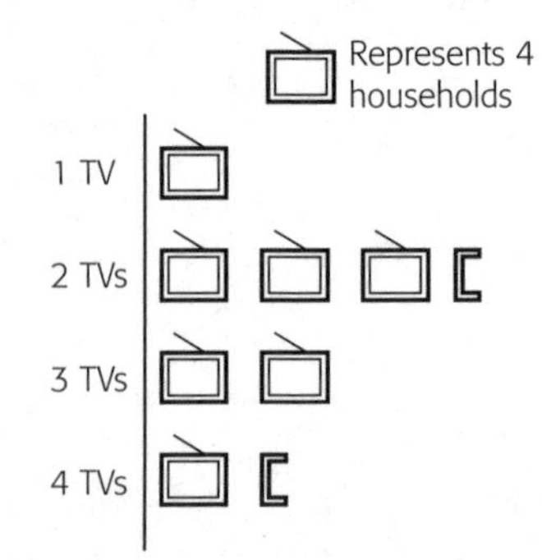

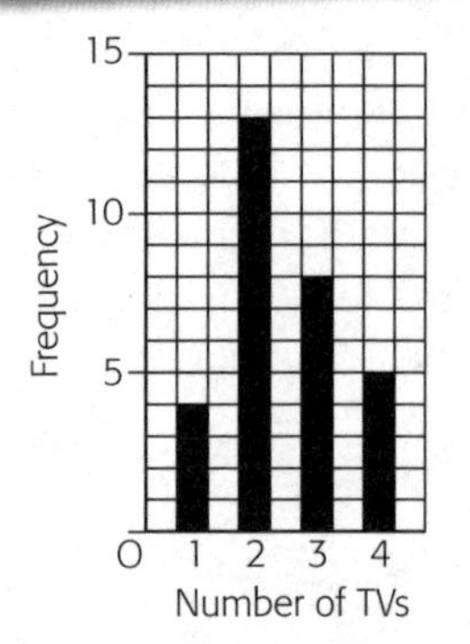

A **vertical line chart** is used for discrete data; it is like a bar chart but with lines instead of bars.

In a **pie chart**, the whole circle (360°) represents the total frequency. Each sector slice shows the **proportion** for each group.

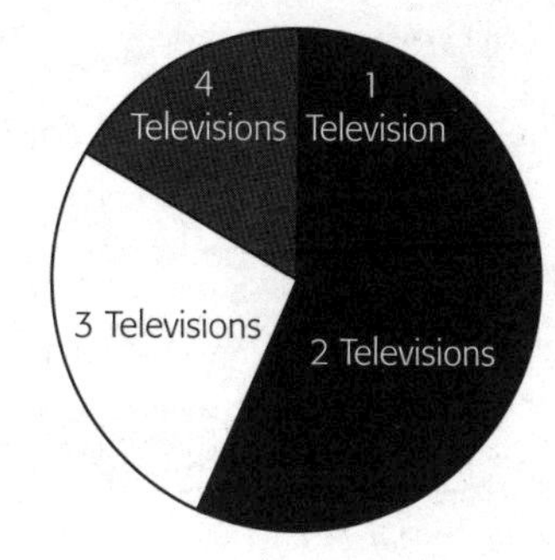

To find the angle for each sector, work out:

$$\frac{360°}{\text{total frequency}} = \frac{360°}{30} = 12°$$

So one TV needs an angle of 12°.

Then multiply your answer by the group frequency to find the sector angle for each group:

$12° \times 4 = 48°$
$12° \times 13 = 156°$
$12° \times 8 = 96°$
$12° \times 5 = 60°$

Check angles total 360°.

Histograms

To represent **grouped continuous data** you should use a **histogram** or **frequency diagram**. These look like bar charts, but the bars touch and don't have to be equal widths.

Here is some data on how long people took to complete a puzzle:

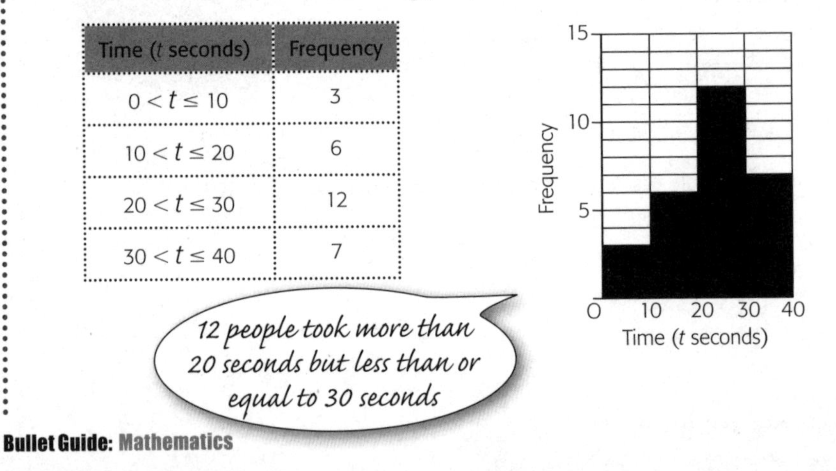

Time (t seconds)	Frequency
$0 < t \le 10$	3
$10 < t \le 20$	6
$20 < t \le 30$	12
$30 < t \le 40$	7

Scatter graphs

These are used for data that comes in pairs, e.g. height/weight, age of car/selling price, time spent revising/exam result.

The graphs show if there is a **correlation** (relationship) between the data.

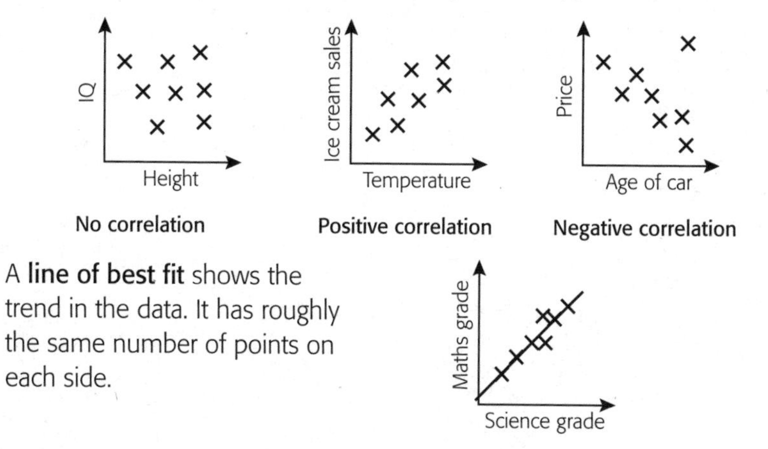

A **line of best fit** shows the trend in the data. It has roughly the same number of points on each side.

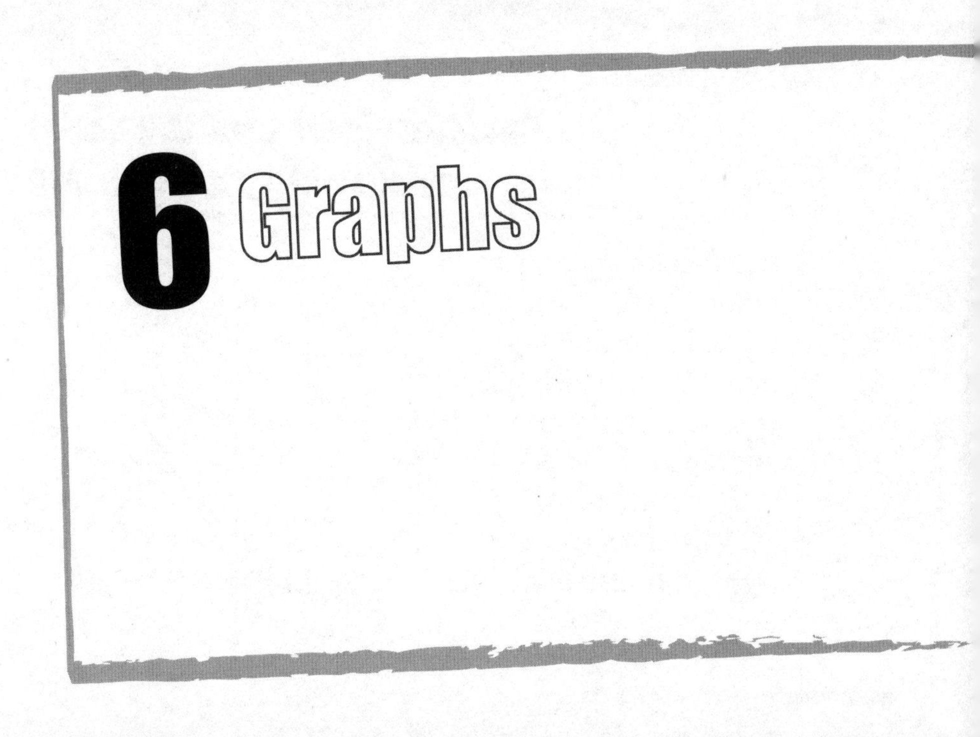

6 Graphs

Coordinates

Coordinates are used to give the position of a point on a grid

- The **origin, O**, is the point (0, 0).
- **Coordinates** are a pair of numbers which tell you how far right/left and up/down a point is from the origin.
- (**3**, **2**) means move **3 right** from the origin and **2 up**.

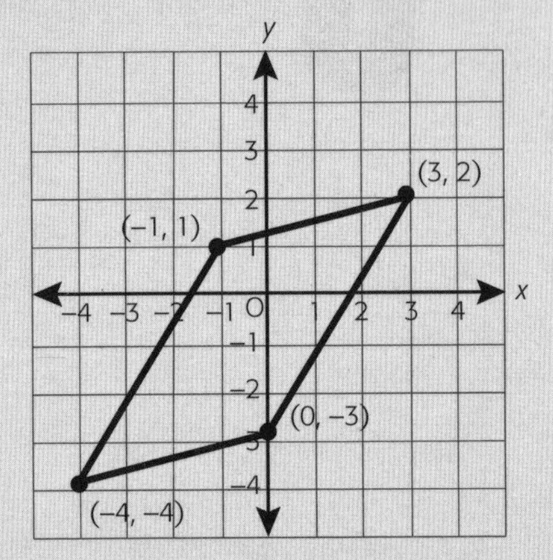

- (−4, −4) means move **4 left** from the origin and **4 down**.
- The **axes** are a pair of lines at right angles to each other.
- The horizontal axis is the ***x*-axis**.
- The vertical axis is the ***y*-axis**.
- Coordinates should always be given in the order (x-coordinate, y-coordinate).
- When you plot coordinates, always start at the origin and go along first and then up/down. You can remember this as 'go along the corridor and then up the stairs'.
- The coordinates of the points on a straight line or a curve follow a pattern.

Lines parallel to the axes

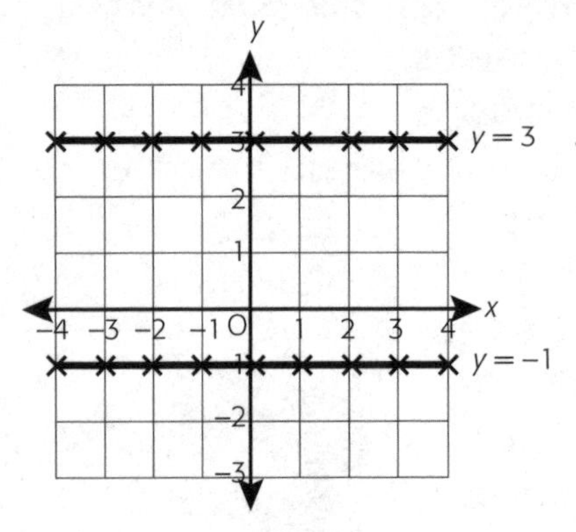

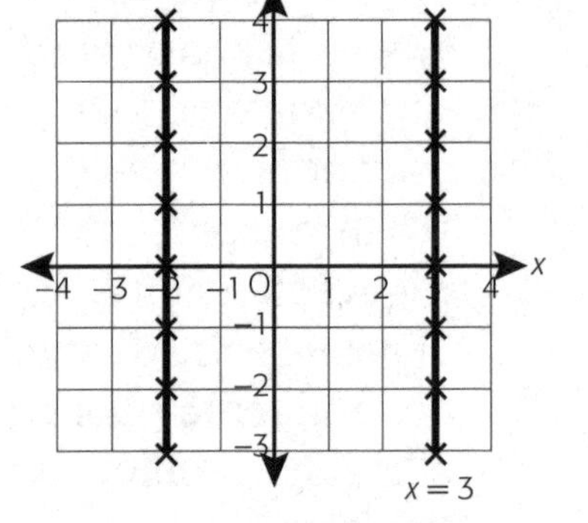

The ***y*-coordinate** of any point on a horizontal line is always the same.

The ***x*-coordinate** of any point on a vertical line is always the same.

Straight-line graphs

To draw the graph of $y = 2x + 1$…

- **Step 1** Make a table of values to help you find the y-coordinate when $x = 0, 1, 2$ and 3.

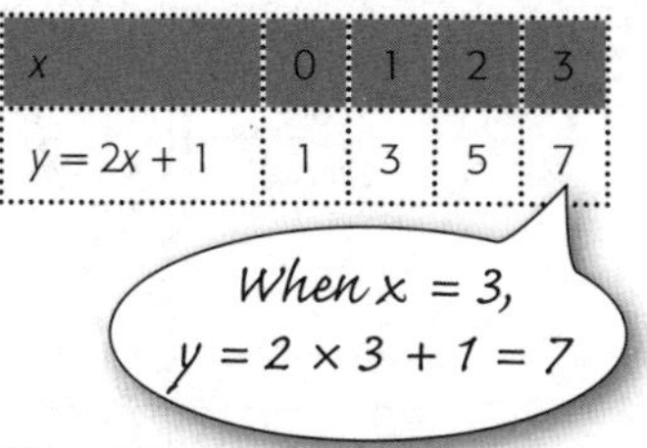

x	0	1	2	3
$y = 2x + 1$	1	3	5	7

- **Step 2** Draw your axes and use pairs of x and y values as the coordinates.
- **Step 3** Extend your line – and don't forget to label it!

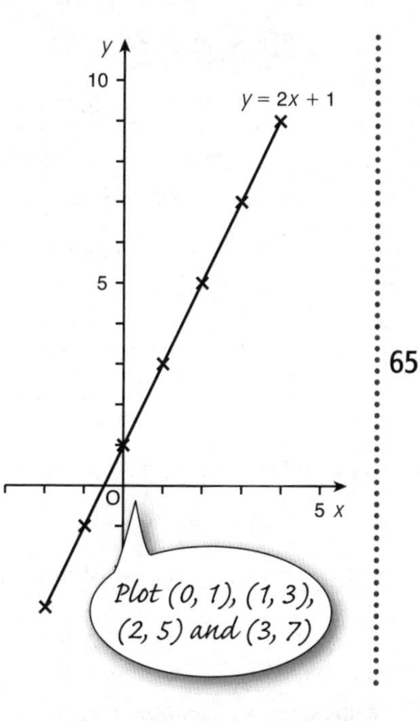

Gradient

All equations of straight lines can be written in the form $y = mx + c$...

- where m is the **gradient** (or slope) *and*
- the coordinates of the ***y*-intercept** are $(0, c)$.

This means you can look at the graph of a straight line and find its equation by:

- working out the gradient $= \dfrac{\text{rise}}{\text{run}}$
- finding where the line cuts the y-axis.

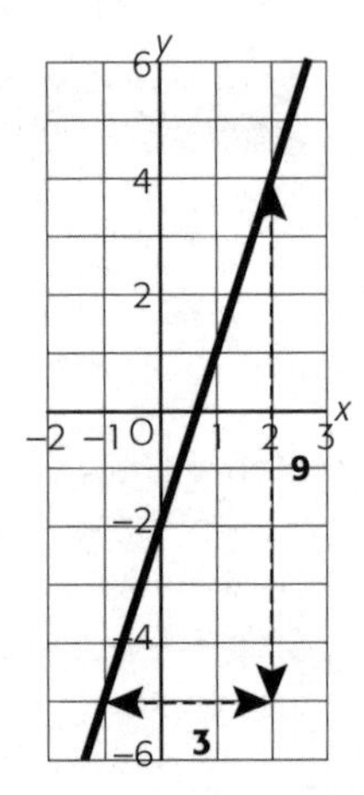

This graph has a gradient of $\frac{9}{3} = 3$

and it cuts the y-axis at $(0, -2)$.

So the equation of the line is $y = 3x - 2$.

You can also know what the graph of a line looks like by looking at its equation:

* The line $y = 4x - 2$ has gradient $= 4$ and y-intercept $(0, -2)$.
* The line $y = -3x + 6$ has gradient $= -3$ and y-intercept $(0, 6)$.

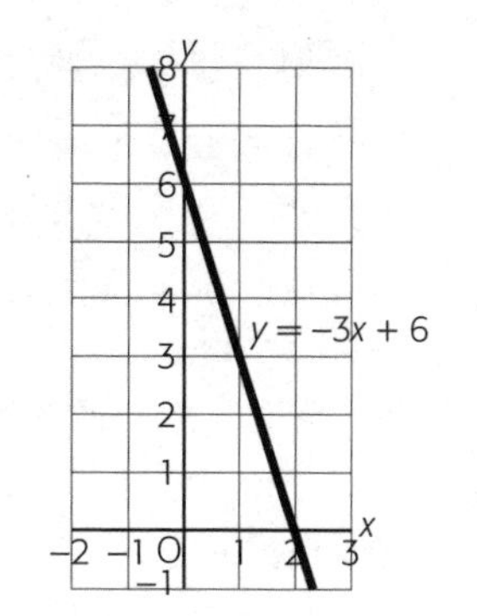

TOP TIP

Uphill slopes are **positive**.
Downhill slopes are **negative**.

Simultaneous equations

Simultaneous equations (which we look at in more detail in Chapter 9) can be solved by drawing graphs:

- draw the graphs of the two equations you need to solve
- read off the coordinates of the point where the lines meet
- write down the coordinates
  $x = \ldots$ and $y = \ldots$

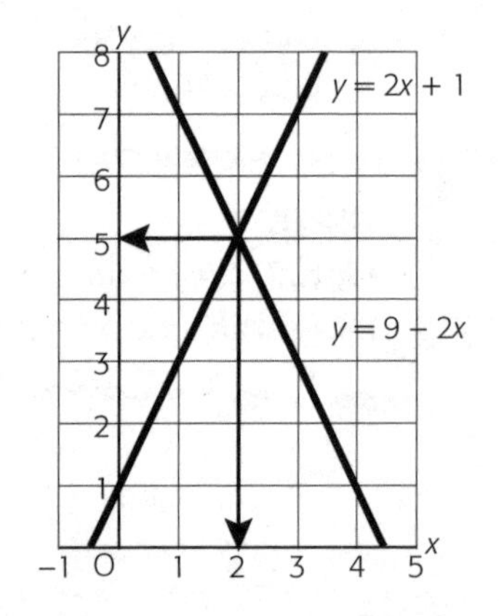

Example

The lines $y = 2x + 1$ and $y = 9 - 2x$ **intersect** (meet) at the point (2, 5). So the solution to the simultaneous equations:

$y = 2x + 1$ and

$y = 9 - 2x$ is

$x = 2$ and $y = 5$

Curves

In a **quadratic equation** (see Chapter 9) the highest power of x is x^2. The graph of a quadratic equation is called a **parabola**. You can make a table of values to plot quadratic curves.

x	−3	−2	−1	0	1	2	3
$y = x^2 + 1$	10	5	2	1	2	5	10

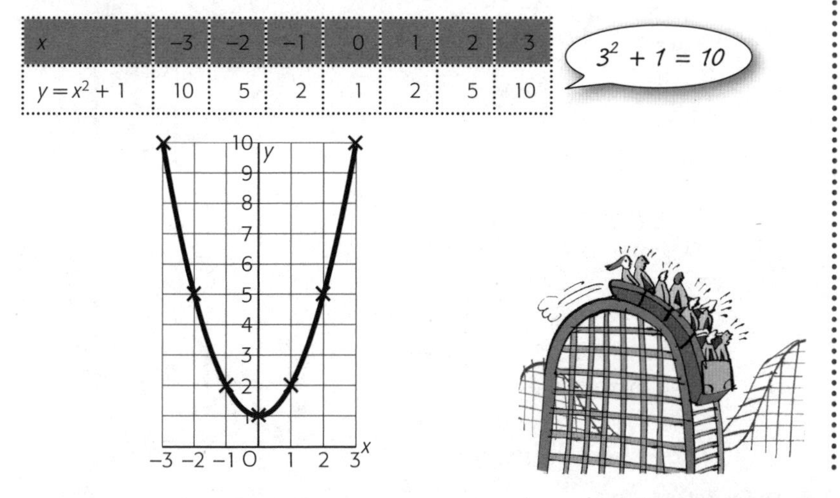

Describing transformations

1 **Translation** – shape slides from one position to another. B is a translation of A by the **vector** $\begin{pmatrix} 3 \\ -2 \end{pmatrix}$ (or 3 right and 2 down).

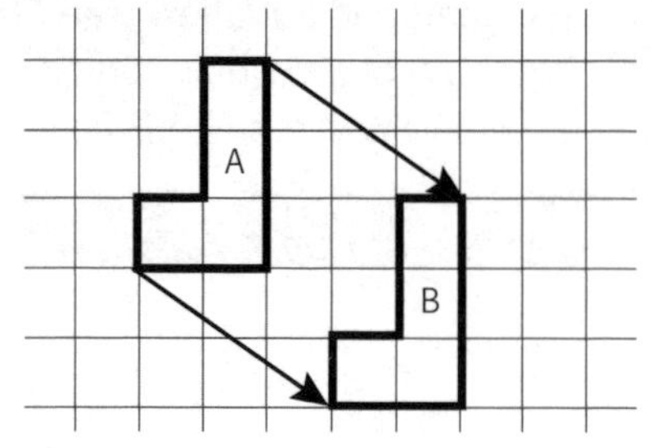

2 **Enlargement** – shape changes size. B is an enlargement of A by **scale factor** 2 (it's twice as big) about centre (0, 0). **Ray lines** drawn through the corners of the shapes meet at the **centre of enlargement**.

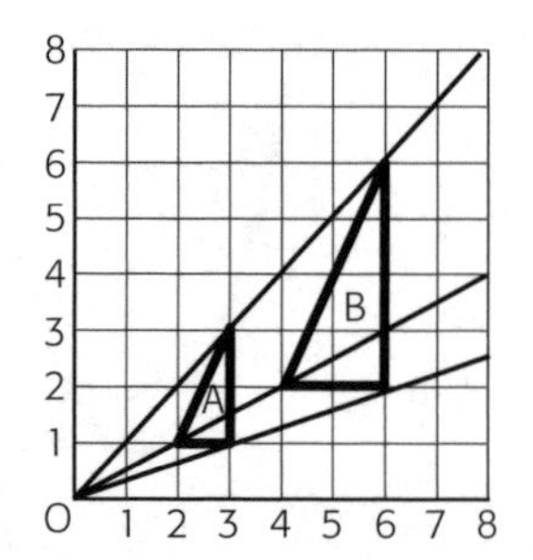

3 **Rotation** – shape turns around. B is a rotation of A through 90° clockwise about the origin. You must state the **angle**, **direction** and **centre of rotation**.

4 **Reflection** – shape flips over. B is a reflection of A **in the line** $x = 3$. The mirror line is halfway between the two shapes.

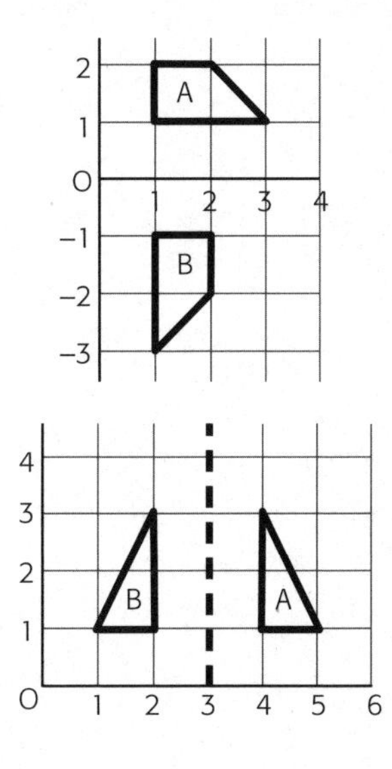

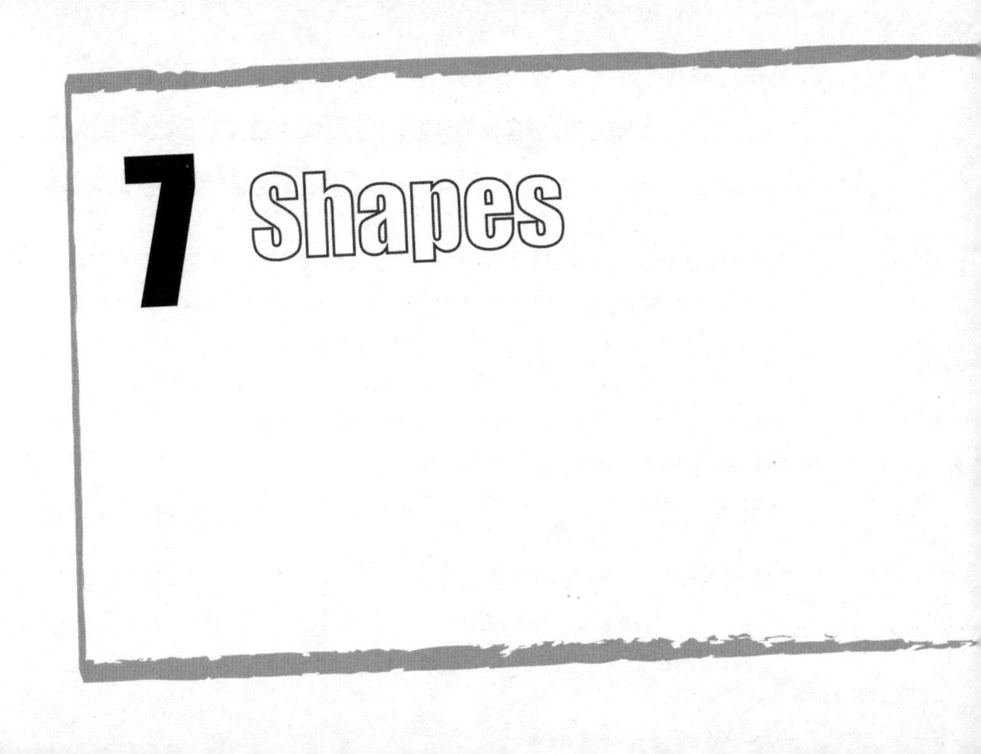

7 Shapes

Some useful words

Part of geometry is the study of shapes and their properties

* **2D shapes** are **flat** shapes. 2D shapes are classified by how many straight edges they have (e.g. a quadrilateral is any four-sided shape) and their symmetry. Many different types of triangle and quadrilateral have special names.
* **3D shapes** are **solid** shapes which have length, width and height. A **polyhedron** is any 3D shape whose faces are all polygons (e.g. triangles, quadrilaterals, pentagons, etc.).

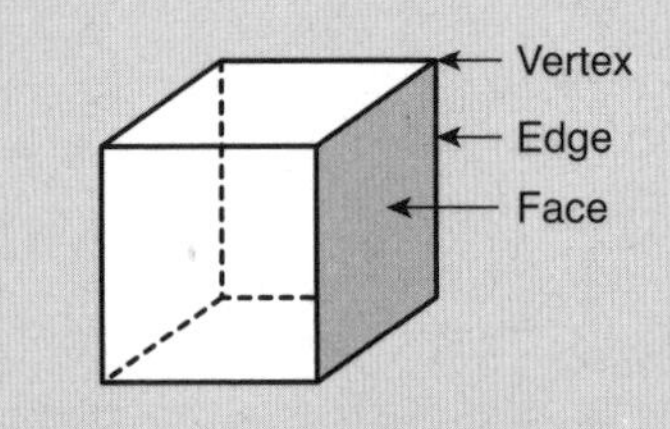

* **Congruent shapes** are exactly the same shape and size.
* **Similar shapes** are enlargements of each other. They are the same shape but different sizes.
* A **net** is a flat shape that can be folded up to make a 3D shape.
* **Perimeter** is the distance all the way round a shape.

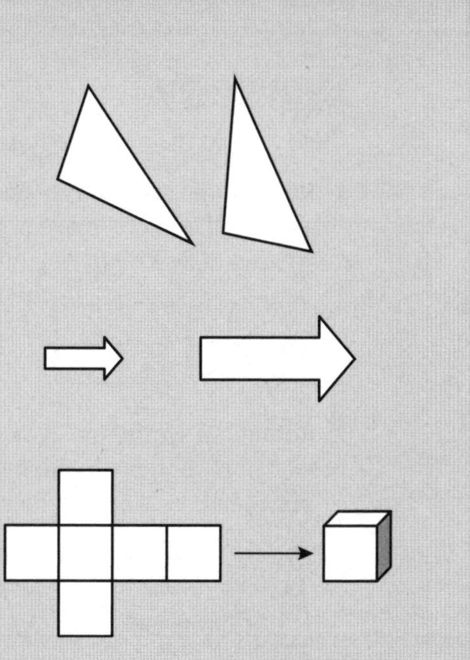

Symmetry

There are two main types of symmetry:

1 **Reflectional symmetry** – one half of the shape is the mirror image of the other half. You can draw a mirror line on the shape.

2 **Rotational symmetry** – when you rotate a shape, its order of rotational symmetry is the number of times it fits into its original position in one full turn. These shapes have rotational symmetry of…

…order 1

…order 2

…order 5

Naming triangles

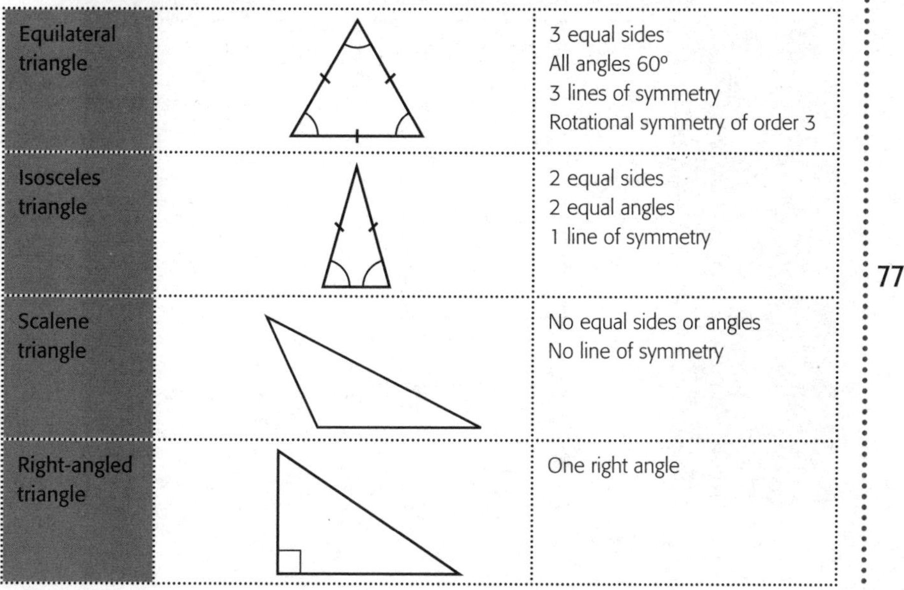

Equilateral triangle		3 equal sides All angles 60º 3 lines of symmetry Rotational symmetry of order 3
Isosceles triangle		2 equal sides 2 equal angles 1 line of symmetry
Scalene triangle		No equal sides or angles No line of symmetry
Right-angled triangle		One right angle

Naming quadrilaterals

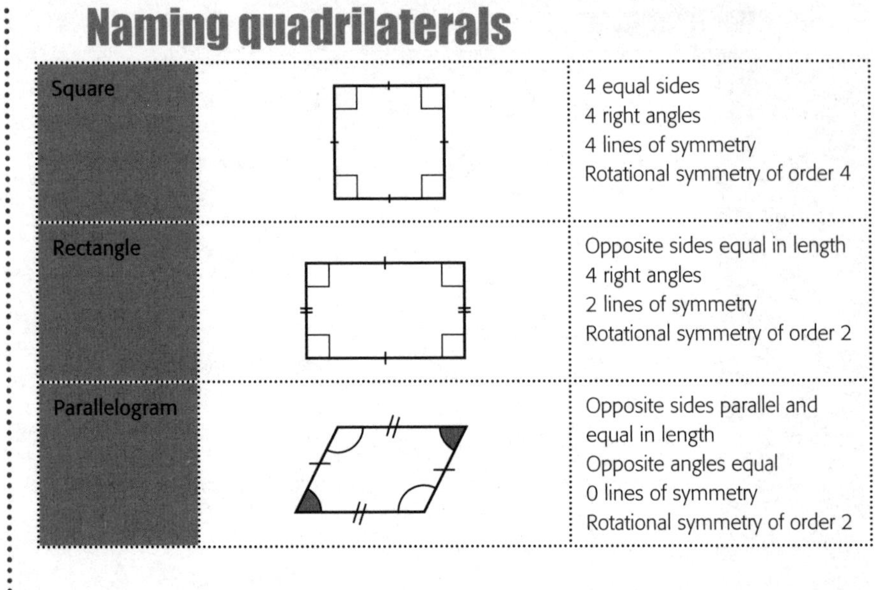

Name	Shape	Properties
Square		4 equal sides 4 right angles 4 lines of symmetry Rotational symmetry of order 4
Rectangle		Opposite sides equal in length 4 right angles 2 lines of symmetry Rotational symmetry of order 2
Parallelogram		Opposite sides parallel and equal in length Opposite angles equal 0 lines of symmetry Rotational symmetry of order 2

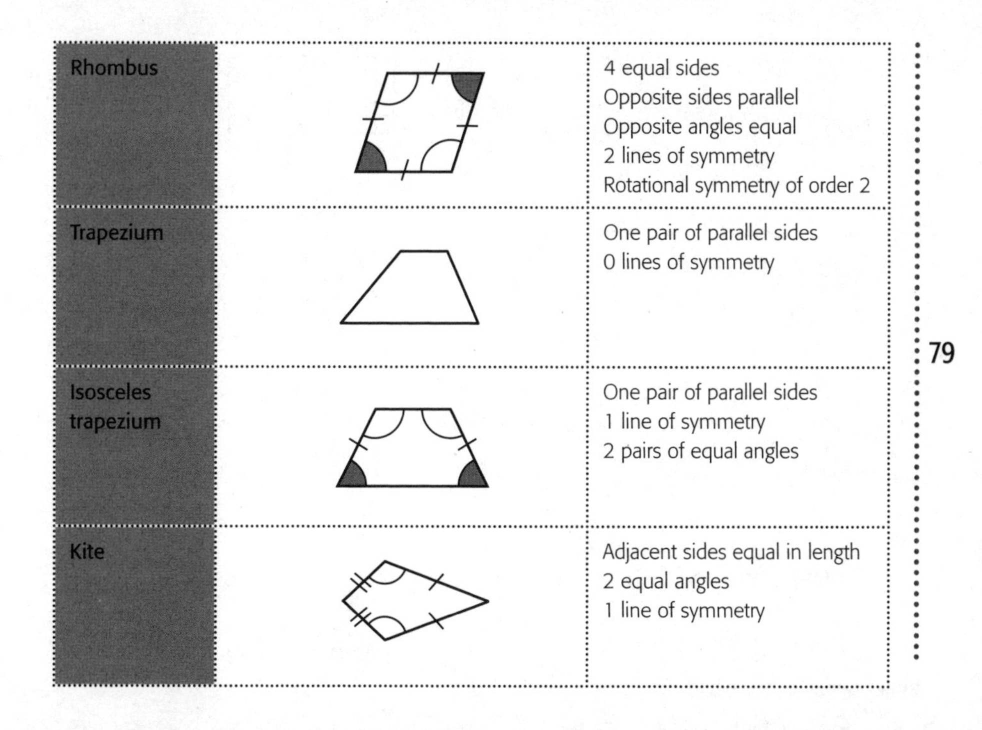

Rhombus		4 equal sides Opposite sides parallel Opposite angles equal 2 lines of symmetry Rotational symmetry of order 2
Trapezium		One pair of parallel sides 0 lines of symmetry
Isosceles trapezium		One pair of parallel sides 1 line of symmetry 2 pairs of equal angles
Kite		Adjacent sides equal in length 2 equal angles 1 line of symmetry

Area

Area of a rectangle = **l**ength × **w**idth

$A = lw$

Area of a parallelogram = **b**ase × **h**eight

$A = bh$

Area of a triangle = half **b**ase × **h**eight

$A = \frac{1}{2} b \times h$

Area of a trapezium $= \frac{1}{2}(a + b) \times h$

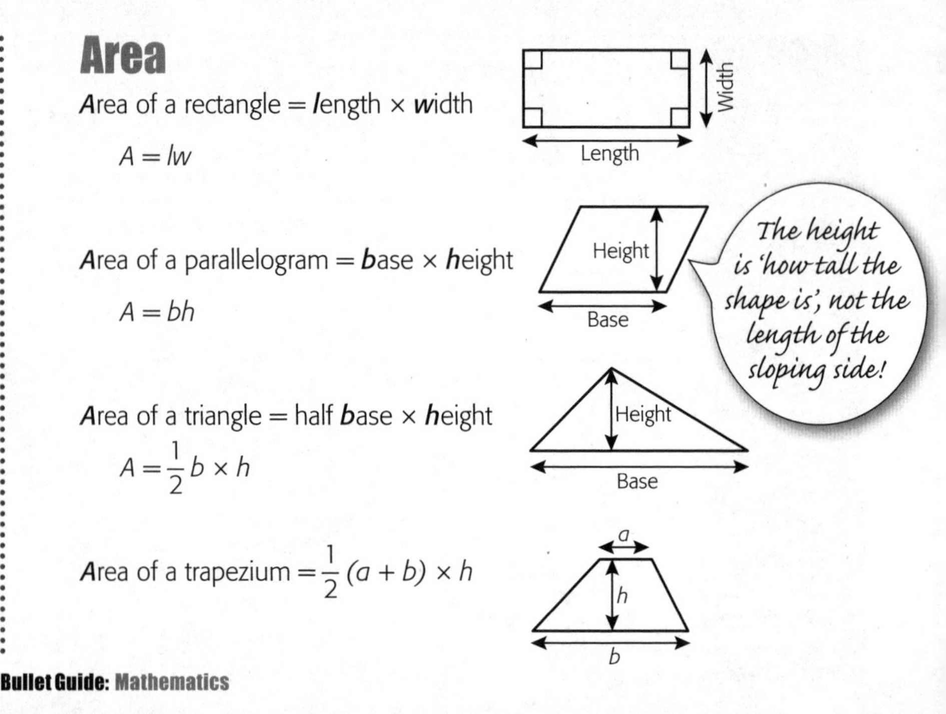

Circles

- π (pi) is the number 3.141592654…
  It is an **irrational number** which means the decimal part carries on for ever and never repeats. You can use 3.14 as an approximation for π.

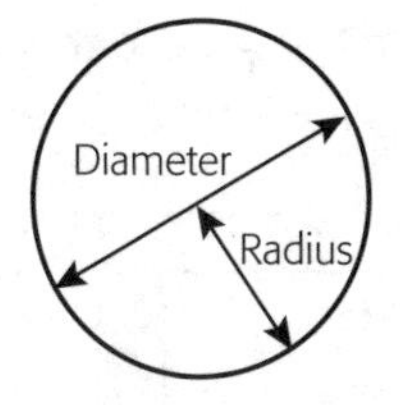

- Diameter of a circle = 2 × radius.
- The **circumference** is the distance around the circle.
  Circumference = π × diameter or 2π × radius.
  In symbols, $C = \pi d$ or $C = 2\pi r$.
- Area of a circle = πr^2 (pi × radius squared).

Example

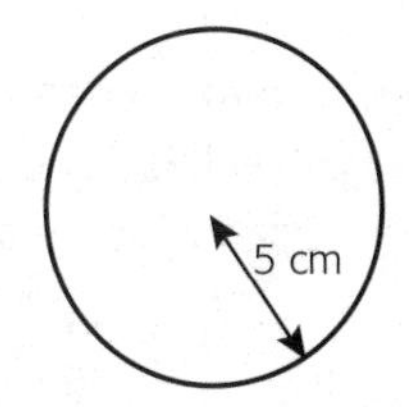

$$\text{Circumference} = 2\pi r$$
$$= 2 \times 3.14 \times 5 = 31.4 \text{ cm}$$
$$\text{Area} = \pi r^2$$
$$= 3.14 \times 5^2 = 3.14 \times 25$$
$$= 78.5 \text{ cm}^2$$

3D shapes

A **prism** is a 3D shape with two identical end faces connected by rectangles (or parallelograms). A prism has a **constant cross-section** which means it is the same shape all the way through.

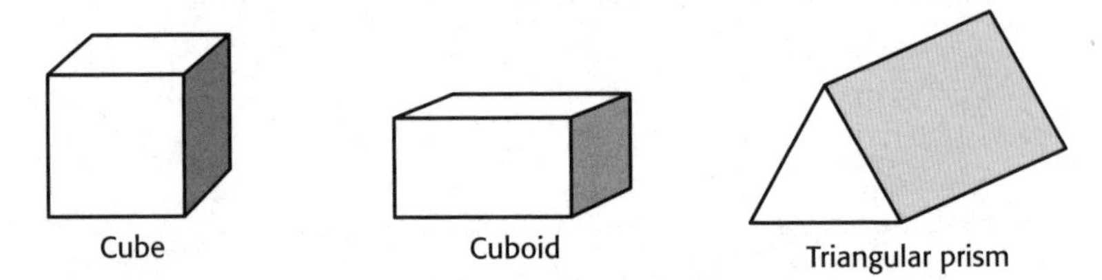

* Volume of a prism = area of cross-section × length.

* A **pyramid** has a polygon for a base and triangular sides which meet at a point.

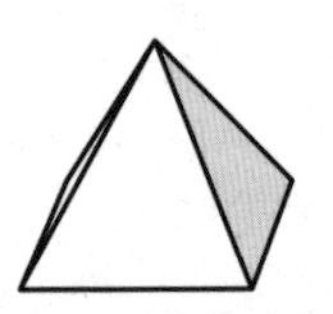

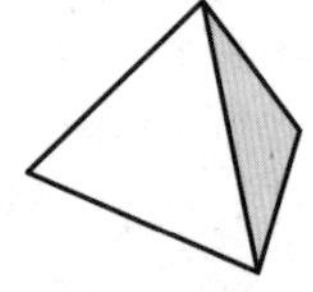

* The **surface area** of a 3D shape is found by adding up the area of each face.

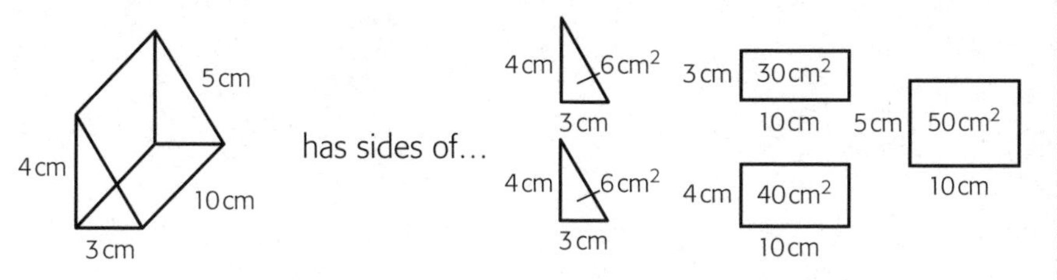

So surface area = 6 + 6 + 30 + 40 + 50 = 132 cm²

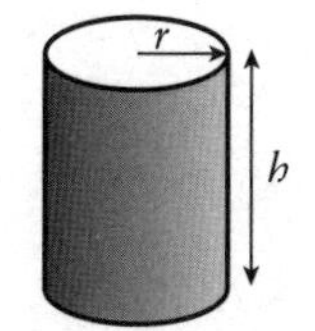

* Volume of a cylinder = $\pi r^2 h$

  Surface area of a cylinder = $2\pi r^2 + 2\pi rh$

* A sphere is a ball shape.

  Volume of a sphere = $\frac{4}{3}\pi r^3$

  Surface area of sphere = $4\pi r^2$

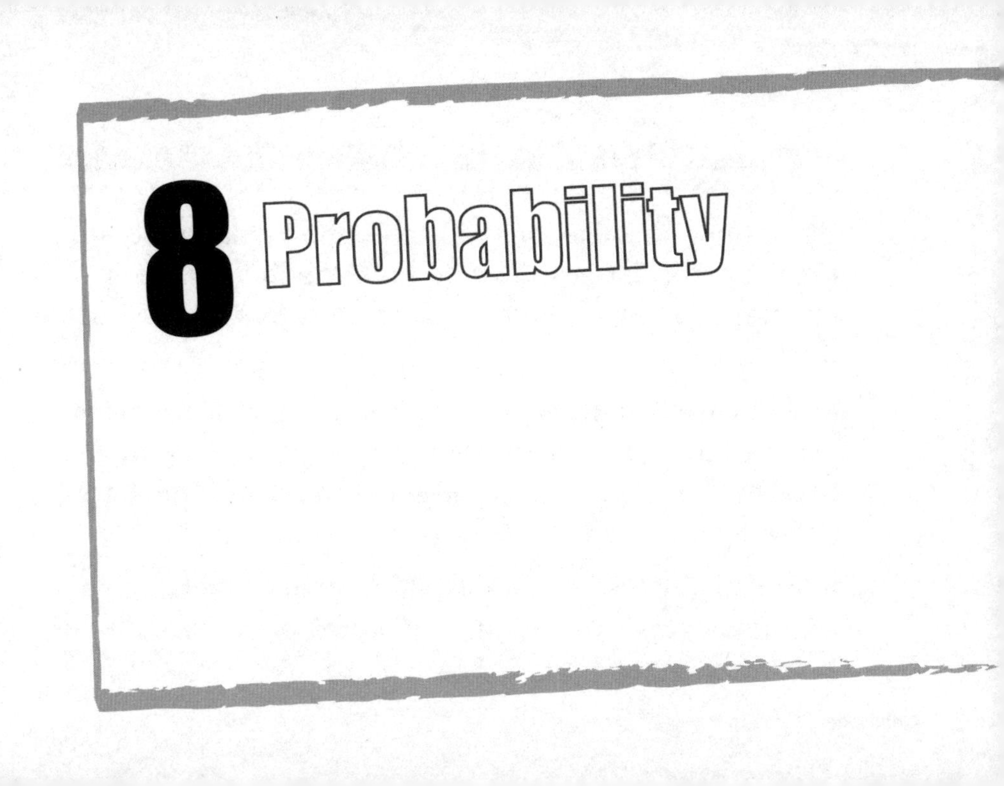

8 Probability

Chance

Probability is about the mathematics of chance

Life is unpredictable, but you can often work out how likely a particular event would be. **Probability theory** enables you to work out how likely it is that you will win the lottery.

Probability is often much misunderstood; unlikely events occur all the time and are only remarkable when they have been predicted, after all someone wins the lottery every week – it is just unlikely to be you!

In probability theory you talk about **experiments** (such as throwing a dice) and the **outcomes** of the experiment or **events** (getting a 1, 2, 3, 4, 5 or 6).

You can write a probability as a decimal, a percentage or a fraction.

- **Random** events don't follow a set pattern or rule.
- **Certain events** have a probability of **1** or **100%**.
- **Impossible events** have a probability of **0** or **0%**.
- An event that has an equal chance of happening or not happening has a probability of 0.5 or ½ or 50%.

You can record probabilities on a probability scale:

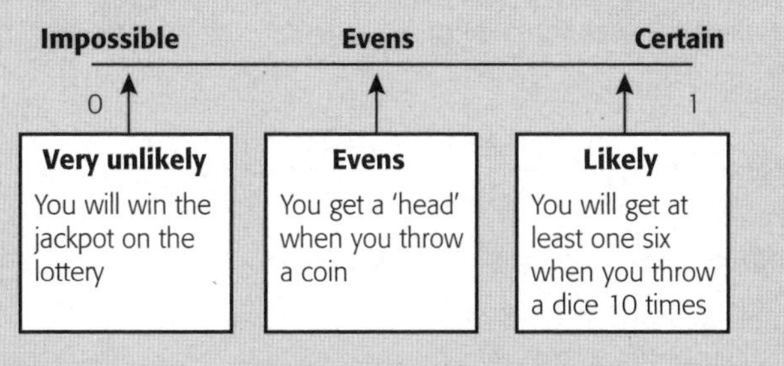

Biased and unbiased processes

When all outcomes of a process are **equally likely**, the process is **fair** or **unbiased**. A fair coin is equally likely to land 'heads' as 'tails'.

When some outcomes of a process are more likely to occur than others, then the process is **biased** or **unfair**. For example, a biased dice might be more likely to land on a six.

When all outcomes are equally likely, you can work out the probability of a particular outcome using:

$$\text{Probability} = \frac{\text{number of favourable outcomes}}{\text{total number of possible outcomes}}$$

A shorthand way of writing 'the probability that event A will happen' is P(A).

Example

A bag contains 5 red balls, 8 yellow balls and 3 blue balls.
A ball is taken from the bag at random. What is the probability that it is a red ball?

Solution: $P(\text{red ball}) = \dfrac{\text{number of red balls}}{\text{total number of balls}} = \dfrac{5}{16}$

Example

When you throw a dice, what is the probability it lands on an even number?

Solution: There are 6 numbers on a dice, 3 of which are even.

So: $P(\text{even}) = \dfrac{3}{6} = \dfrac{1}{2}$ or 0.5

Probability calculations

Any event, A, either happens or doesn't happen so…

P(A happens) + P(A does not happen) = 1

and P(A does not happen) = 1 – P(A happens)

Example

The probability that Alan is late for work is 0.15.

Solution: P(Alan is on time) = 1 – P(Alan is late)
= 1 – 0.15 = 0.85

Expected frequency

Expected frequency = number of trials × probablity of success

If you threw a coin 1,000 times, you would expect to get around 500 'heads'.

Tree diagrams

These are used to solve problems. You **multiply** probabilities along the branches and **add** the probabilities at the ends of the branches.

Example

When Andy plays tennis the probability he wins a game is 0.8. Andy plays two games. What is the probability he wins exactly one game?

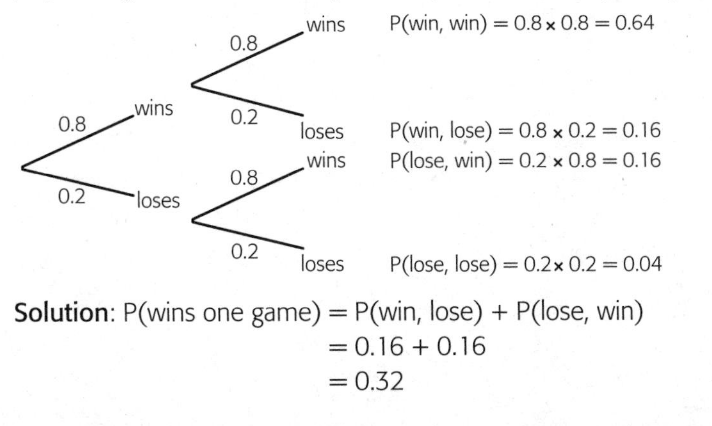

Solution: P(wins one game) = P(win, lose) + P(lose, win)

= 0.16 + 0.16

= 0.32

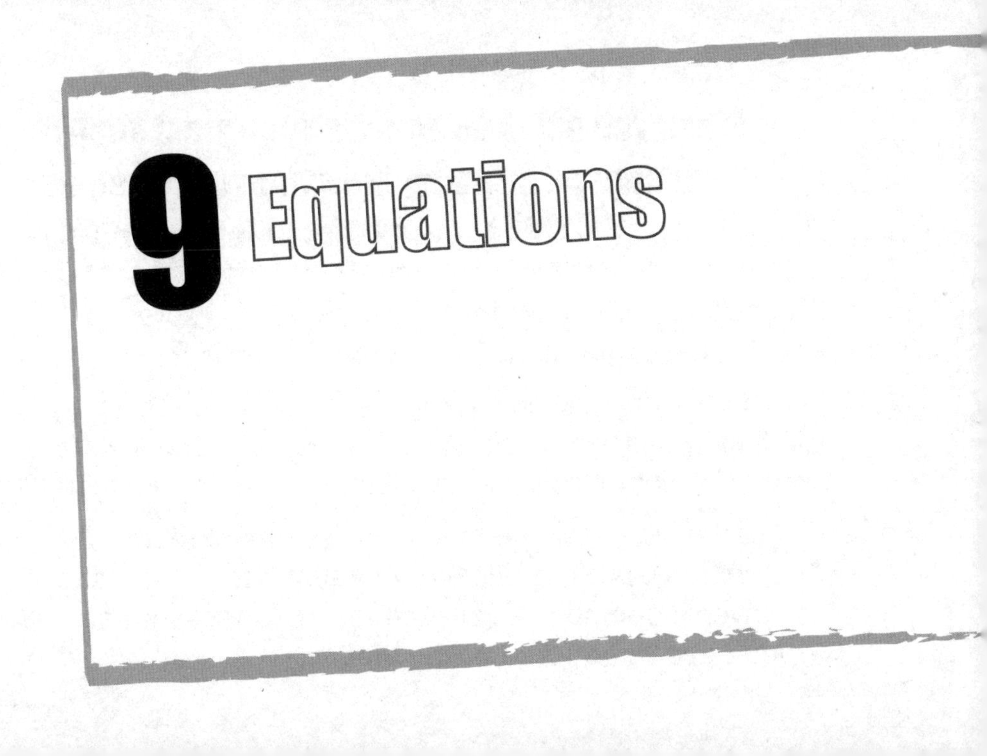

9 Equations

Keeping your balance

When you solve an equation you must keep it balanced by doing the same to both sides – this is called the balance method

An **equation** is a statement which says 'this expression' equals 'that expression'. For example, $4 + 5 = 9$.

Sometimes an equation has missing numbers which you can find by **solving** the equation. For example, $\square + 3 = 10$. Usually **letter symbols** are used for these missing numbers.

You can solve simple equations by looking at them (**inspection**), but to solve harder equations you need to use inverse operations. For example, to 'undo' subtract 5, you need to add 5 to **both sides**.

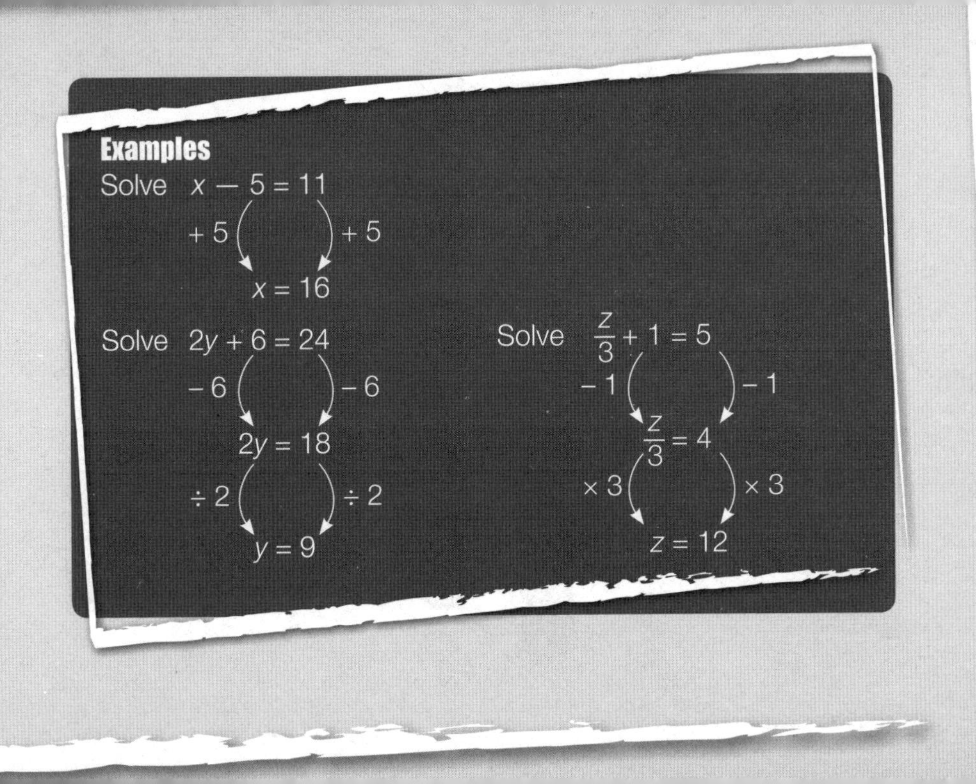

Examples
Solve $x - 5 = 11$
$+5$ $+5$
$x = 16$
Solve $2y + 6 = 24$
-6 -6
$2y = 18$
$\div 2$ $\div 2$
$y = 9$
Solve $\frac{z}{3} + 1 = 5$
-1 -1
$\frac{z}{3} = 4$
$\times 3$ $\times 3$
$z = 12$

Unknowns on both sides

- When an equation has an unknown on **both sides**, you need to get all the unknowns on one side first.
- When an equation has brackets, expand the brackets first. For example: $3(x - 2) = 14$ means $3x - 6 = 14$.

Inequalities

- The symbol $>$ means 'is greater than'. The symbol $<$ means 'is less than'. For example, $10 > 7$ and $4 < 9$.
- $2 \leq n < 5$ means any number (n) that is **greater** than or **equal** to 2 but **less than** 5.
- You can solve inequalities in the same way as you solve equations. Just keep the inequality sign.

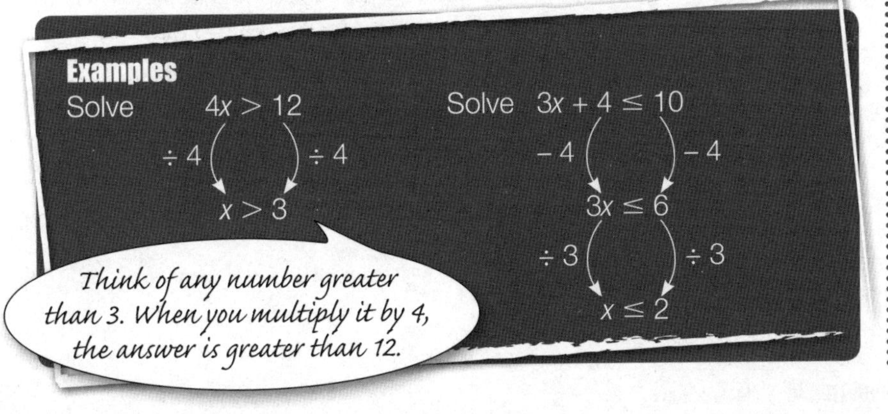

Solving simultaneous equations

This is where you are a given a pair of equations to solve, like:

$x = 8$ and $y = 2$ or
$x = 13$ and $y = -3$, etc.

$$x + y = 10$$

$$x - y = 2$$

$x = 10$ and $y = 8$ or
$x = 56$ and $y = 54$, etc.

Lots of solutions work for each equation (in fact, an infinite number), but only one pair of solutions works for **both** equations **simultaneously**.

When you have the same number of x's or y's, you can solve the equations by adding or subtracting them to get rid of x or y.

$$\begin{array}{rl} & x + y = 10 \\ + & x - y = 2 \\ \hline & 2x \quad = 12\text{, so } x = 6 \end{array}$$

As $x + y = 10$ and $x = 6$, then $y = 4$.

OR

$$\begin{array}{rl} & x + y = 10 \\ - & x - y = 2 \\ \hline & 2y = 8\text{, so } y = 4 \end{array}$$

As $x + y = 10$ and $y = 4$, then $x = 6$.

Check: $4 + 6 = 10$ ✔ and $6 - 4 = 2$ ✔

Sometimes you haven't got the same number of x's or y's. When this happens you can multiply **all the terms** in one or both equations to make them match.

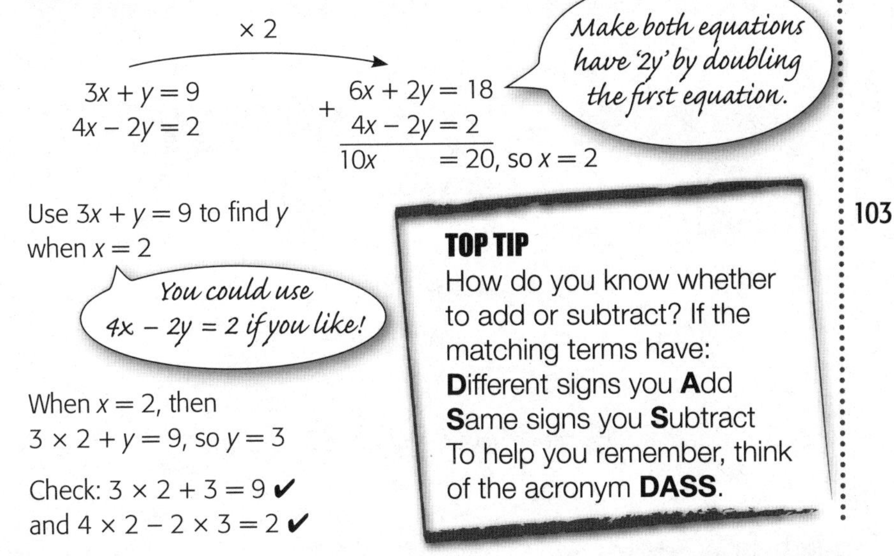

$\times 2$

$$3x + y = 9 \qquad\qquad 6x + 2y = 18$$
$$4x - 2y = 2 \qquad + \quad 4x - 2y = 2$$
$$10x \quad = 20, \text{ so } x = 2$$

Use $3x + y = 9$ to find y when $x = 2$

When $x = 2$, then
$3 \times 2 + y = 9$, so $y = 3$

Check: $3 \times 2 + 3 = 9$ ✔
and $4 \times 2 - 2 \times 3 = 2$ ✔

TOP TIP

How do you know whether to add or subtract? If the matching terms have:
Different signs you **A**dd
Same signs you **S**ubtract
To help you remember, think of the acronym **DASS**.

Quadratics

* A **quadratic** is an expression in which the highest power is x^2.
* When you expand a pair of brackets you can end up with a quadratic.

Expand $(2x + 3)(x - 5)$

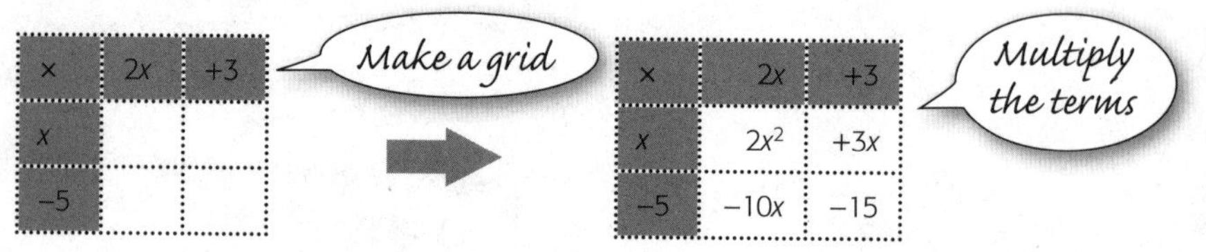

×	$2x$	$+3$
x	$2x^2$	$+3x$
-5	$-10x$	-15

Now write down the terms inside the grid:

$$(2x + 3)(x - 5) = 2x^2 + 3x - 10x - 15$$

$$= 2x^2 - 7x - 15$$

* Quadratic equations have 0, 1 or 2 real solutions.

You can solve some quadratic equations by factorizing. This means you rewrite the equation using two brackets.

The numbers that go in the brackets must...

1 $x^2 + 7x + 12 = 0$
$(x + 3)(x + 4) = 0$
Either $x + 3 = 0$ or $x + 4 = 0$
So $x = -3$ or -4

...multiply to give +12 and add to give +7. Use +3 and +4.

2 $x^2 + 4x - 5 = 0$
$(x + 5)(x - 1) = 0$
So $x = -5$ or 1

...multiply to give −5 and add to give +4. Use +5 and −1.

3 $x^2 - 6x + 8 = 0$
$(x - 4)(x - 2) = 0$
So $x = 4$ or 2

...multiply to give +8 and add to give −6. Use −4 and −2.

Solving quadratics using the formula

You can solve a quadratic equation in the form $ax^2 + bx + c = 0$ by using the formula:

$$x = \frac{-b \pm \sqrt{b^2 - 4ac}}{2a}$$

Solve $2x^2 + 3x - 5 = 0$

Plugging $a = 2$, $b = 3$ and $c = -5$ into the quadratic formula gives:

$$x = \frac{-3 \pm \sqrt{3^2 - 4 \times 2 \times -5}}{2 \times 2} = \frac{-3 \pm \sqrt{9 + 40}}{4} = \frac{-3 \pm \sqrt{49}}{4} = \frac{-3 \pm 7}{4}$$

So $x = \frac{-3 + 7}{4} = \frac{4}{4} = 1$ or $x = \frac{-3 - 7}{4} = \frac{-10}{4} = -2.5$

Trial and improvement

Solving equations **using trial and improvement** means making a **sensible guess** at the solution and then **improving** on your guess. Here's how to solve $x^3 + 5 = 25$ by trial and improvement:

Try $x = \ldots$	$x^3 + 5$	Compare with 25
2	$2^3 + 5 = 13$	too small
3	$3^3 + 5 = 32$	too big
2.5	20.625	too small
2.8	26.952	too big
2.7	24.683	too small
2.75	25.79…	too big

The answer is between 2 and 3

The answer is between 2.5 and 2.8

The answer is between 2.7 and 2.75, so $x = 2.7$ to 1 d.p.

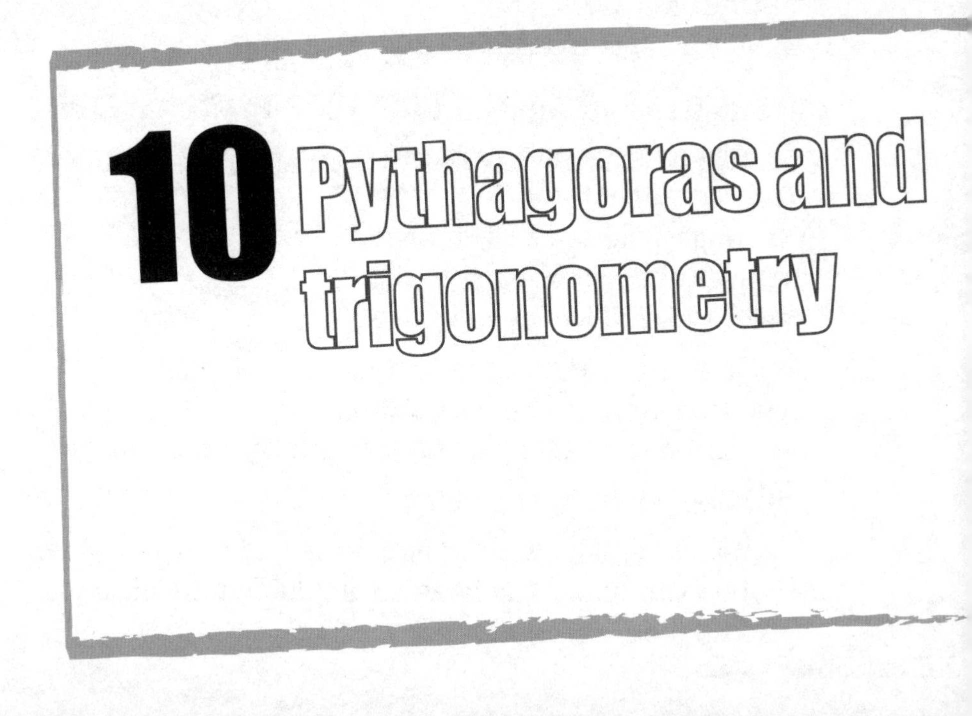

10 Pythagoras and trigonometry

Pythagoras' theorem

You can use Pythagoras' theorem and trigonometry to solve problems involving right-angled triangles

The longest side of a right-angled triangle (opposite the right angle) is called the **hypotenuse**.

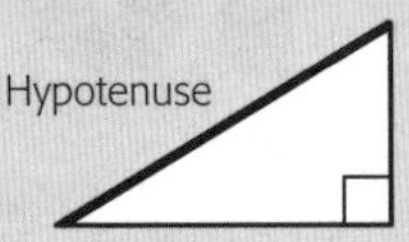

You should use Pythagoras' theorem when you know **2 sides** of a right-angled triangle and are asked for the **3rd side**.

You should use trigonometry:

- when you know **2 sides** and are asked for an **angle** *or*
- when you know **1 side** and **1 angle** and are asked for a **2nd side**.

Pythagoras' theorem states that for a right-angled triangle:

$$a^2 + b^2 = c^2$$

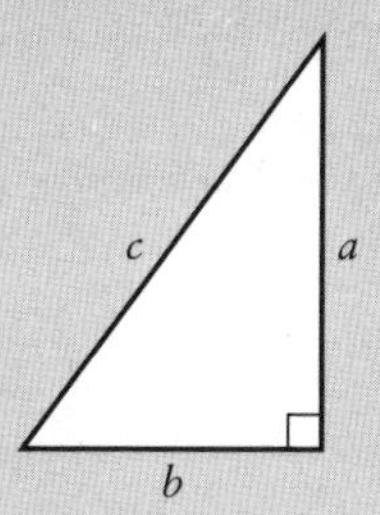

Pythagoras' theorem *only* works for right-angled triangles so you can use it to check whether a triangle is right-angled.

Does $8^2 + 6^2$ equal 10^2?
Yes, $64 + 36 = 100$

So this triangle *is* right-angled.

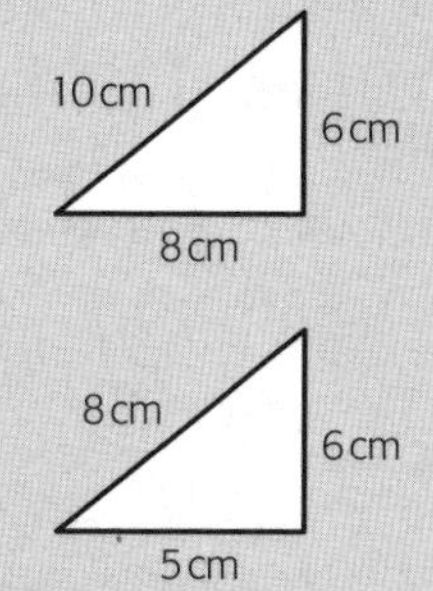

Does $5^2 + 6^2$ equal 8^2?
No, $25 + 36 = 61$ not 64

So this triangle *is not* right-angled.

Finding the hypotenuse (longest side)

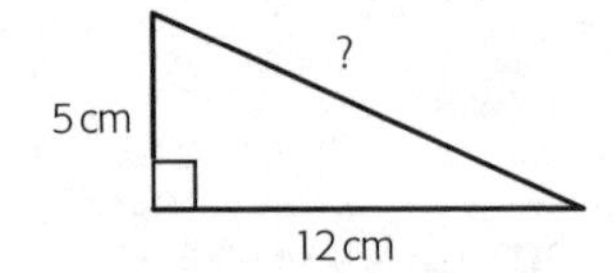

$5^2 = 25$ and $12^2 = 144$

Step 1: Square the 2 sides you know

$$
\begin{aligned}
c^2 &= a^2 + b^2 \\
&= 25 + 144 \\
&= 169
\end{aligned}
$$

Step 2: Add together the squares

$$
\begin{aligned}
\text{So } c &= \sqrt{169} \\
&= 13 \text{ cm}
\end{aligned}
$$

Step 3: Square root the result

Finding one of the shorter sides

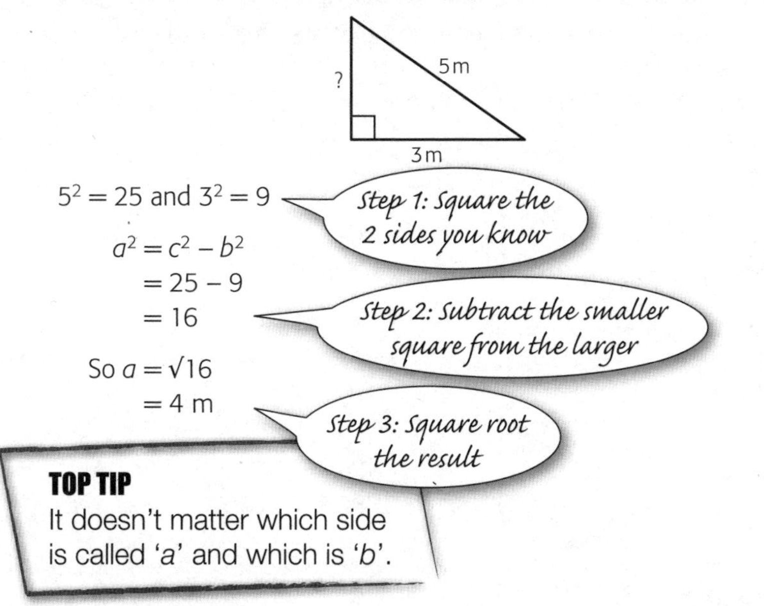

$5^2 = 25$ and $3^2 = 9$

$$a^2 = c^2 - b^2$$
$$= 25 - 9$$
$$= 16$$

So $a = \sqrt{16}$
$$= 4 \text{ m}$$

TOP TIP

It doesn't matter which side is called '*a*' and which is '*b*'.

Labelling triangles

- The letter x or the Greek letter θ (theta) are often used for an unknown angle.
- The two shorter sides of a right-angled triangle are called the **opposite** and the **adjacent**.

θ Opposite

The opposite is opposite the angle, θ.

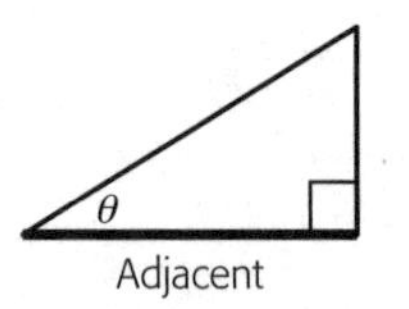

The adjacent is next to the angle, θ.

Trigonometric functions

You can use the trigonometric functions – sine (sin), cosine (cos) and tangent (tan) – to help you find missing sides or angles in a right-angled triangle:

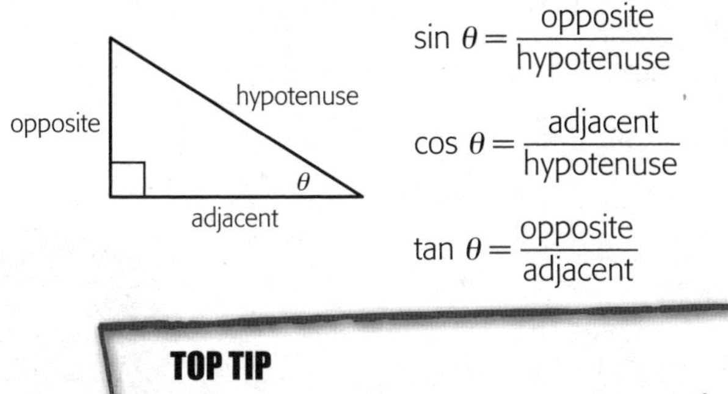

$$\sin\theta = \frac{\text{opposite}}{\text{hypotenuse}}$$

$$\cos\theta = \frac{\text{adjacent}}{\text{hypotenuse}}$$

$$\tan\theta = \frac{\text{opposite}}{\text{adjacent}}$$

TOP TIP

You can use the word *'soh-cah-toa'* (pronounced phonetically) or a phrase like '**S**illy **O**ld **H**arry, **C**aught **A** **H**erring, **T**rawling **O**ff **A**merica' to help you remember these.

Solving problems

To find the missing side.

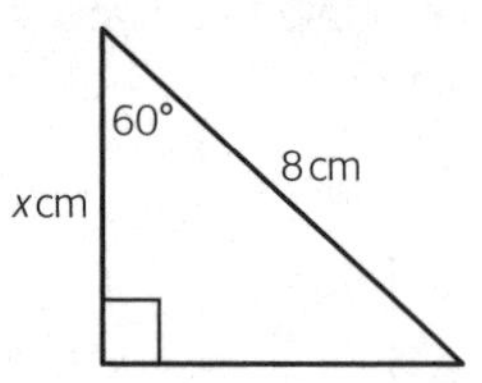

You know:

- an angle, 60°
- and the hypotenuse, 8 cm

and you want the adjacent, x.

The part of *'soh-cah-toa'* which involves *'a'* and *'h'* is *'cah'* so use:

$$\cos\theta = \frac{\text{adjacent}}{\text{hypotenuse}}$$

Substitute in the values to get: $\cos 60° = \frac{x}{8}$

Rearrange: $x = \cos 60° \times 8$

Now $\cos 60° = 0.5$ and so $x = 0.5 \times 8$

$x = 4$ cm

To find the value of θ.

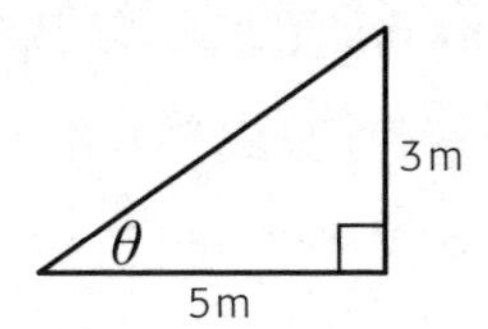

You know:

* the opposite, 3 m
* and the adjacent, 5 m

and you want the angle, θ.

The part of *'soh-cah-toa'* which involves *'o'* and *'a'* is *'toa'* so use:

$$\tan\theta = \frac{\text{opposite}}{\text{adjacent}}$$

Substitute in the values to get: $\tan\theta = \frac{3}{5} = 0.6$

Use the $\tan^{-1}$ (inverse tan or arctan) button on your calculator to find the value of θ.

$$\theta = 31.0° \text{ to 1 d.p.}$$

Trigonometry for triangles without right angles

You can work out the area of any triangle using:

$$\text{Area} = \frac{1}{2}ab \sin C$$

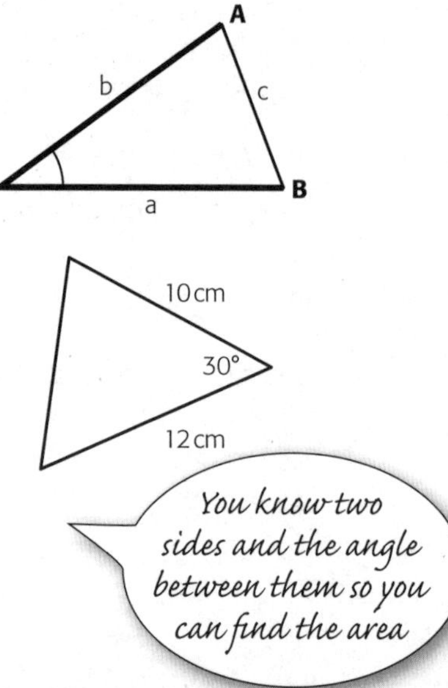

Example

Find the area of this triangle:

$$\begin{aligned}\text{Area} &= \frac{1}{2} \times 10 \times 12 \times \sin 30^\circ \\ &= 60 \times \sin 30^\circ \\ &= 60 \times \frac{1}{2} \\ &= 30 \text{ cm}^2\end{aligned}$$

You can use the **sine** rule and the **cosine** rule to work out sides or angles of any triangle.

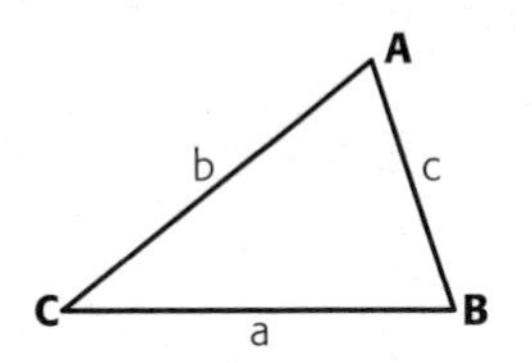

The cosine rule is $a^2 = b^2 + c^2 - 2bc \cos A$

The sine rule is $\dfrac{\sin A}{a} = \dfrac{\sin B}{b} = \dfrac{\sin C}{c}$

Use the cosine rule when you know:

* 3 sides and want to find an angle
* 2 sides and the angle between them and need the 3rd side.

Use the sine rule when you know:

* 2 sides and 1 angle and want a 2nd angle
* 1 side and 2 angles and want a 2nd side.

Further reading

* Eastaway, Rob & Askew, Mike, *Maths for Mums and Dads* (Square Peg, 2010). An interesting guide to the latest methods taught in schools. A must for any parent!
* Goldie, Sophie & Smith, Alan, *Foundation GCSE Mathematics for Edexcel* (Hodder Education, 2006). Lots of examples and further questions to work through.
* Johnson, Trevor & Neill, Hugh, *Complete Mathematics* (Hodder Education, 2010). An informative guide to maths, which includes lots of examples and practice questions.
* Lawler, Graham, *Understanding Maths: Basic Mathematics Explained* (Studymates, 1999). An accessible book which gives you a step-by-step guide to maths.
* Potter, Lawrence, *Mathematics Minus Fear* (Marion Boyars Publishers, 2006). An entertaining book that takes away the fear of maths.